A Field Guide to the Subterranean

Also by Justin Hocking

The Great Floodgates of the Wonderworld: A Memoir
P.S.: The Wolves

A Field Guide to the Subterranean

Reclaiming the
Deep Earth and
Our Deepest Selves

A Memoir

Justin Hocking

A FIELD GUIDE TO THE SUBTERRANEAN

This is a work of creative nonfiction. The author reconstructed various scenes and dialogue from memory. Some names and identifying details of individuals and organizations have been changed to protect their privacy; certain time sequences have been altered to heighten the narrative.

Copyright © 2025 by Justin Hocking

All rights reserved under domestic and international copyright. Outside of fair use (such as quoting within a book review), no part of this publication may be reproduced, stored in a retrieval system, or transmitted in any form or by any means, electronic, mechanical, photocopying, recording, or otherwise, without the written permission of the publisher. For permissions, please contact the publisher.

First Counterpoint edition: 2025

Library of Congress Cataloging-in-Publication Data
Names: Hocking, Justin, author.
Title: A field guide to the subterranean : reclaiming the deep earth and our deepest selves : a memoir / Justin Hocking.
Other titles: Reclaiming the deep earth and our deepest selves : a memoir
Description: First Counterpoint edition. | San Francisco : Counterpoint, 2025.
Identifiers: LCCN 2024061414 | ISBN 9781640097018 (hardcover) | ISBN 9781640097025 (ebook)
Subjects: LCSH: Hocking, Justin. | Young men—Colorado—Biography. | Underground areas—Colorado—Anecdotes. | Mining—Colorado—Anecdotes. | Outdoor life—Colorado—Anecdotes. | Coming of age.
Classification: LCC CT226 .H635 2025 | DDC 978.8 [B]—dc23/eng/20250220
LC record available at https://lccn.loc.gov/2024061414

Jacket design by Farjana Yasmin
Jacket photograph © iStock / Sebastien Clozel
Book design by Laura Berry
Watercolor shapes © Veris Studio / Adobe Stock
Quartz © lunata / Adobe Stock

COUNTERPOINT
Los Angeles and San Francisco, CA
www.counterpointpress.com

Printed in the United States of America

1 3 5 7 9 10 8 6 4 2

*For Roger Hocking (1940–2017)
and Barry Lopez (1945–2020)*

Reclamation: *n.* 1: The act of making a claim or protest 2: The act or process of restoring land that has been mined to a more natural state 3: To recover possession of something lost

Contents

Part One   Subterranean   1

Part Two   Heights   83

Part Three   Equatorial   149

Acknowledgments   207

Part One

Subterranean

Hot Springs, Age Eleven

PUSHING OUT THE HEAVY DOOR, THE DECEMBER night chill stings his bare skin alive, his tiny nipples. This small Colorado town's ever-present aura of chlorine and sulfur light up his sinuses. Rippling and glowing emerald beneath him: a million gallons of mineral-rich water in a cement basin longer than a football field.

Columns of steam spin off the surface, spiraling forty feet into the sky, a forest of vapor.

Sharp crunch of rock salt under his heels as he hobbles (*no running at the pool*) across frosted cement to the smaller, hotter of the two pools. The earth-heated water pincushions his skin as he submerges—feet, shins, knees, thighs, and tender parts—pausing at the belly button before the full-body dunk.

He finds his mother soaking with her friends on the stairs. Her lovely skin, hair streaked with feldspar. Mist candles from her fingertips when she waves, gifts him with her radiant smile. Glasses off, she squints a little. While she mingles he slips away to the larger pool.

Doggy-paddling through the shallow end, other bathers materialize from the fog—steam emanating from their head-crowns, their collarbones—then disappear. Beads of creamy copper light play on the water's surface, reflecting from antique lampposts.

Halfway to the deep end a vapor column shifts, unveiling a teenage couple kissing against the marble pool wall. Most eleven-year-olds would continue side-stroking toward the diving boards, oblivious. But he's already learned to tread the waters of adult lust. He pretends not to stare, looks away, then back again, glimpses the edge of the girl's tongue, until she detects him in her peripheral vision. Smiling now as she kisses, the smile for her boyfriend but also somehow for the boy, almost naked as he is with them in the same vapor understory, the same warm mineral bath. Watching opens a secret shame—the cavernous wound he keeps hidden since an older neighbor began strip-mining his innocence, four years earlier. But also something new and humming, a honey hive in his stomach.

Up above the red mountains, the moon briefly sheds its cloud-robe. Then disappears again.

This is when it falls.

A snowflake wide as his thumb knuckle. Then another, and another, the passing storm's silvery gauze-work gently dressing his shame, swaying his attention back to the sensual night as it braids a frost-crown through his hair. Each snowflake's mirror image rises up from below; they meet on the water's surface before dissolving.
    The ephemeral kiss of hot and cold, light and shadow, eros and thanatos. Underworld and sky.

How to Use This Book

MY MOTHER BROUGHT ME TO THE HOT SPRINGS Pool on weeknights; on weekends before he moved away to California, my father drove me to ghost towns, marble quarries, abandoned gold mines, the Vapor Caves. Followed by regular visits to another of my favorite places: the High Country Gems and Minerals shop in Glenwood Springs, Colorado. The owner, Lee, wore black military-issue glasses and a buzzed haircut. Always the rhythmic fizz and clicks of a rotary rock tumbler whirring away in the back room. The dusty, mineral smell of the place and the wood-paneled walls gave High Country the feel of a cabin or a geological research station.

What always drew my attention first: a little basket of triangular objects with a handwritten sign—*Rocky Mountain Shark Teeth, 75¢ Each.*

"How is that possible?" I asked Lee.

"Same reason you find these," he said, pointing to fossilized skeletons of tiny fish and seashells. He then explained that our whole valley was submerged in a shallow sea for thousands of years.

As I would later learn, *glossopetrae* is the scientific term for fossilized shark teeth. In his *Natural History*, Roman naturalist Pliny the Elder shares, with some skepticism, the widely held belief that they fell from the sky during lunar eclipses. In the seventeenth century a court physician to the Medici family matched the glossopetrae he found in the hills above Florence with those from an actual shark's jaw; this unsettled the concept of a static earth, just as I was astounded by the image of prehistoric sharks swimming slow circles above my hometown.

For sale in other baskets in Lee's shop: arrowheads, chunks of milky quartz, halved purple geodes, lapidary jasper. All arranged atop a glass display case with a gray push-button that operated a miniature chain-driven Ferris wheel. I loved watching Lee's treasures scroll around inside the display: lapis lazuli, gold flakes in plastic vials, silver and turquoise jewelry, chunks of malachite, deep-blue azurite. Pressing that gray button elevatored my consciousness down into the wondrous sub-terra, where heat, pressure, and time transform ordinary elements into shimmering minerals and prismatic gems—and where the planet worked its alchemical magic on me too.

Years later, as a teenager in San Diego, I fell for the ocean. But as a young boy I crushed hard on the earth.

After my father paid Lee one dollar for a pocket-sized copy of a Golden Nature Guide to *Rocks and Minerals*, featuring four hundred full-color illustrations, my earth-crush grew more hopeless. The cover alone was mesmerizing: an illustration of a rock hammer, surrounded by a gorgeous quartz cluster and various maroon and aquamarine gemstones.

"Nothing is as important in the natural world as our own earth and the rocks beneath our feet," reads the first sentence of the foreword.

In the next chapter, author Herbert S. Zim claims that "increased use of mineral resources is a direct measure of progress, and that it is impossible to think of a time when dependence on minerals will cease." Now, as an adult, I'm struck by Zim's extractive ideology, the notion that *nothing*—not children or aspen trees or polar bears or the original human inhabitants of North America—matches the importance of mineral wealth. Of course that didn't quite register for me as a smitten seven- or eight-year-old.

"Take this book on hikes, trips and vacations," reads a passage from another prefatory chapter called "How to Use This Book."

I brought the Golden Nature Guide with me everywhere, in my school backpack or my external-frame camping pack. The back pocket of my Levi's or on the dashboard of my father's old pickup. I was less interested in the actual text than marveling, page after page, at the '50s-era illustrations of amethysts and gleaming emeralds. Somehow it endured multiple moves: to California for high school (where I hid it deep in a drawer, along with my Golden Field Guide to *Birds of North America*), bundled up in boxes to college back in Colorado, then across the country to Brooklyn, and ultimately west again to Oregon.

"This book will help you understand the importance of rocks and minerals in our daily lives. As you make observations and collect specimens, check your book or make marginal notes. In the long run, your records may be as important as the specimens you collect."

I didn't manage to save a single rock or shark tooth gifted to me by my late father. But I still have the Golden Nature Guide, with its waxy pages and wrinkled spine. It's the only book I've held on to for the better part of forty years. Thumbing through it now soothes the father-shaped hole that tore through me after his death in 2017. I'm not the dedicated rock hound he was, yet the act of assaying the past—digging and sieving through the loose rubble of memory, weighing and testing for value, keeping obsessive written records—has become my vocation.

The idea of explaining to readers how to use any book feels antiquated. The epitome of mid-century American paternalism. Please do what you will—sift randomly through these sections, landing on whatever glimmers catch your eye. Or keep your finger on the gray button, watching the narrative unfurl start to finish. Above all, please read with care. The hard labor of unearthing the past rewards us with glittering delights, though the descents—and what we uncover in the light-starved depths—are not without certain dangers.

Groundshock, 1973

IT BEGINS AS A LOW ROLLING RUMBLE, LIKE A TRAIN off in the distance.

My father sits on a wooden bench outside the ranch house, pulling on cowboy boots, whistling Johnny Cash and Waylon Jennings tunes. Colorado sun warm on his cheeks, his square chin.

The rumbling stirs the horses; the dogs whimper and cower beneath the porch, like when a wolf stalked the property, months before.

Seconds later it reaches him like a minor earthquake, like subsurface thunder.

My father stands.

My mother's inside, holding her pregnant belly while sitting on the edge of the bed, which by now rocks violently, nearly throwing her to the floor. She calls out my father's name, "Roger?"

"Roger" can be used as an affirmative, as in *Roger that*. My mother's usage is just the opposite.

On screen: a small, earthy mound rises from the ground, a slow-motion birth. A male voice with a smoker's timbre and crisp enunciation narrates the documentary film, shot in the 1960s as a propaganda campaign for Project Plowshare. "To perform a multitude of peaceful tasks for the betterment of mankind," the narrator explains, "man is exploring a source of enormous, potentially useful energy: the nuclear explosion."

On screen, the mound expands into a steep, swollen hill.

Plowshare: an agricultural implement for turning soil, for gouging a cleft in the earth. The archetypal idea of the plowshare derives from the bible, in which the prophet Isaiah urges the people to beat their swords into plowshares.

On screen, the hill grows into a mountain, and then the mountain—which is not a mountain—erupts, a hulking black mass edged with feathery projectiles, a pyrotechnic finale of soil and shale.

Under the auspices of Project Plowshare, scientists, engineers, miners, and military personnel detonated twenty-seven nuclear warheads in subterranean shafts of various depths across the continental United States. The project lasted seventeen years, from 1965 until 1973—the year of my birth. "For the benefit of *all nations*," the narrator explains.

Lewis L. Strauss, chairman of the United States Atomic Energy Commission, supervised Plowshare. Strauss was the man responsible for stripping J. Robert Oppenheimer, coinventor of the atom bomb, of his security clearance. In Strauss's opinion, Oppenheimer's increasing concerns about nuclear proliferation and radioactive wastes were pedantic, alarmist.

Strauss and his Plowshare operatives had colossal ambitions. They envisioned large-scale nuclear excavation and quarrying: canals, railroad cuts through mountains, dam construction. They hoped to blast a harbor-sized hole in Cape Thompson, Alaska, and hack a channel through an oceanic reef in the Marshall Islands, all to enhance commerce and transportation.

Their grandest plan: to supplement or even replace the Panama Canal. Four possible routes across Central America were selected and studied. In the Plowshare film, a crude animation illustrates a zipper of nuclear blasts from the Atlantic to the Pacific, to excavate a sea-level channel up to two hundred feet wide and one thousand feet deep, an oceanic superhighway across Costa Rica or Nicaragua. Here are Le Corbusier's and Robert Moses's visions for unimpeded industrial transportation writ large—a freeway in the sea, a tainted slash across the American continent.

Another Plowshare application involved underground engineering and extraction of natural gas and petroleum. *Nuclear fracking*, in other words. Operation Rio Blanco was designed for this purpose, to frack a vast pocket of natural gas, just thirty miles from my parents' house.

The Christian concept of *hell* has origins in the New Testament, where it's described in the Greek as *hades* or with the Hebrew word *gehenna*. The actual gehenna was the "Valley of Hinnom," a garbage heap in a low ravine on Jerusalem's outskirts. A place where refuse was burned. Bodies of suicides were thrown in, to smolder with the muck and char. Authors of the New Testament use gehenna as a metaphor for the final place of punishment for the damned. A linear narrative: hell as perpetual destination, endpoint, a fiery conclusion. The place where Dante cast endless sinners, upside down in hellholes of their own carving, or in the tangled forest of suicides, or with the fratricides in the ninth circle, all of them eternally condemned.

Our Western inheritance, then: the concept of the deep underground as wasteland, dump, terminus of the unredeemable.

Western Colorado towns named for their mineral wealth: Crystal, Marble, Redstone, Gypsum, Silverton, Silt, Copper Mountain, Leadville. New Castle, Colorado, was named after New Castle, England—a historic coal-producing city. On my father's side, Hocking ancestors originated from Cornwall, England; some were fishermen, but most made their living as coal or tin miners, in the epicenter of industrialized mining.

Four years after they married, my parents bought a small ranch outside Silt, across the Colorado River and south of New Castle. My father wasn't a miner himself, but as a civil engineer he made structural plans for tunnels and sewers—underground work. And like generations of Hockings before him, he pursued mineral wealth. In his case, the early 1970s boom in the oil shale industry on the Western Slope attracted him. He performed surveys for infrastructure around local oil operations; he engineered sprawling man-camps to house shale miners. Eventually he launched his own successful firm and named it El Dorado Engineering, after the mythical city of gold that enticed Spanish conquistadors to the new world.

My parents fixed up the modest ranch, mended fences and corrals. They raised horses and an Australian shepherd with one brown and one crystal-blue eye. Regular visitors to the ranch: coyotes, elk, foxes, crows and hawks, and once, a wolf. The wolf kept his distance, but drove the dogs half-mad with fear.

They looked forward to their first child—especially my mother, who worked shifts as a maternity ward nurse at Valley View Hospital. By April of 1973, they'd decided on my name. My mother was a careful parent. Avoided alcohol, ate carrots and lettuce and tomatoes from their garden plot. At the small diner in Silt, she sometimes asked ranchers and miners at neighboring tables to extinguish their cigarettes. This embarrassed my father, a man who prided himself on *not letting things bother him*.

No one had thoroughly informed my parents about the nuclear testing, but among the residents of Silt, Project Plowshare was common knowledge. The Rulison test, in which the government detonated a single forty-three-kiloton nuclear bomb belowground just twenty miles from town, took place four years earlier, without fanfare. The Rio Blanco test of 1973—wherein Plowshare operatives planned to ignite three nuclear warheads a mile beneath the surface—was just another subterranean exploration in an area developed largely for the purposes of mining.

Everyone knew it was coming.

Linear versus cyclical narratives: in some Indigenous cultures, hell is an intermediary period between incarnations or stages. Many have no concept of hell; they view the underground as a place of ritual, transformation, and even genesis. Modern Puebloans—whose forebears may have hunted on the high deserts around the Rio Blanco nuclear test site—believe their ancient ancestors first emerged from the underworld. They commemorate the emergence with a small hole in the floor of their kivas called a sipapu—a symbolic portal to the birthplace of their people.

In the Project Plowshare film, the narrator acknowledges the problem of radiation, and problems other than radiation: the groundshock, the airblasts, the dustcloud. Fundamental distinctions: earth as a womb, or earth as a tomb of eternal punishment. The underground as an origin point, a sacred realm, versus the underground as a site to exploit, carve up, burn. James Watt, secretary of the Interior under Ronald Reagan, was a fundamentalist Christian known for his extractive boasts: "We will mine more, drill more, cut more timber . . ." He believed large-scale strip-mining of the West was acceptable because the Rapture was imminent.

Craig Hayward was eighteen when he witnessed the Rulison nuclear shot. He was the grandson of the rancher who leased the land to Project Plowshare. The U.S. government promised the Hayward family a share in revenue, should the nuclear method prove viable for fracking natural gas.

"When they touched that thing off, I saw shale cliffs crumbling," Hayward said. "After a while, I saw the ground rolling. It was like a wave coming through. Cars were parked there. They were rocking back and forth."

A small group of protesters gathered near the detonation zone. They feared polluted groundwater, contaminated natural gas seeps, radioactive nucleotides with a half-life of a quarter of a million years. As current landowners near the Rulison site still worry, fifty years later, when energy companies drill for natural gas just miles away.

The 1959 film *Hiroshima Mon Amour*, by French director Alain Resnais, explores the entangled nature of personal and historical memory; it's shot in an experimental style that cracks apart the genres of fiction and documentary and smelts them into a ghostly, shimmering amalgam. Jean-Luc Godard described *Hiroshima* as "Faulkner plus Stravinsky." The mostly impressionistic plot centers a French woman visiting Hiroshima to film an anti–nuclear war film—a postmortem attempt at documenting Japan's holocaust. There she has an affair with a Japanese man; both are married, both still traumatized by war. Though she attempts to comprehend the Hiroshima bombing by visiting a museum and interviewing survivors for her film, her Japanese lover tells her repeatedly that she "saw nothing of Hiroshima." Susan Sontag dubbed Resnais's work "the cinema of the inexpressible." Other critics point to *Hiroshima Mon Amour* as the epicenter of postmodernism. It's about the hopeless inability of a museum—or any conventional, linear narrative—to convey the lived experience of such cataclysmic trauma.

But what of the three nuclear bombs buried and detonated near your family's home, two and a half months before your birth? There were no mass casualties, no immediate human casualties at all, so perhaps narrative is up to the task. But wouldn't the explosion cause the story itself to rock and fracture like those crumbled shale cliffs?

The Rio Blanco nuclear shot stimulated a flow of natural gas, as hoped, but the product was contaminated with radioactivity. Unusable. Just as it was in 1969, after the Rulison test.

After twenty-seven nuclear shots, Project Plowshare began losing popular support. Activists deemed the blasts and their aftermath an environmental catastrophe. Among politicians, the idea of using nuclear weapons to build a better Panama Canal was seen as increasingly preposterous.

Seventeen years after Plowshare's genesis, Rio Blanco marked the very last underground detonation. The endpoint of the project's linear narrative.

I want to believe that, during the Rio Blanco test, my mother felt an internal rocking—the child inside her growing restless, uneasy. Perhaps in reaction to her own physiological stress response. Elevated heart rate, adrenaline spike as the ground shook beneath her—the result of a nuclear blast six times as powerful as the Hiroshima bombing.

I was still in the womb; I saw nothing of Project Plowshare. But was I curious and disturbed, even then, by what took place beneath the surface?

Continental Drift

NOT LONG AFTER I WAS BORN, MY PARENTS MOVED from Silt to the larger town of Glenwood Springs. Maybe they wanted to escape all the drilling and oil exploration, the atomic earthquakes. But when I was five, tectonic plates beneath our house began to crack and shudder once again.

Sitting on the edge of my bed, I felt it.

The tremors of rage began in the morning, then quieted for a time, my mother collapsed on the couch. They resumed when my father came home from work and collided again with my mother. The quaking sometimes lasted through the night, my parents' ragged edges grinding and buckling.

One morning, the shaking knocked loose most of my father's belongings—silver toaster, TV, electric razor, stacks of clothing and shoes. They fell into messy piles on the dining room table, then vanished.

Burning Mountain

I WAS SIX OR SEVEN WHEN I FIRST HEARD BURNING Mountain's origin story, riding in the back seat of a dusty gold International Scout with my best friend, Gabe, his family, and their sofa-sized Newfoundland dog. We were heading up to their remote A-frame cabin—a sanctuary to me after my parents' divorce. Gabe and his parents, Sheila and Ron, along with his four younger siblings, were a kind of surrogate family, the Mountain House always animated and bustling, unlike each of my single parents' homes. We zipped down snowy hillsides on sleds or inner tubes and cross-country skied in the winter; in the warmer months we hiked and bird-watched and waded around in the creek. All the while, I witnessed Gabe's emerging talents for rendering what he saw outside into artists' notebooks and canvas.

As we passed through the town of New Castle on the way to the Mountain House, Sheila pointed out patches of dead vegetation on Burning Mountain, tendrils of smoke rising from thermal vents around the arid hillside's base.

She explained that, long ago, a series of mining accidents and explosions ignited the coal seams beneath the mountain.

"A hundred years later," she said, to my astonishment, "that mountain's insides are still on fire."

In *Arctic Dreams*, author Barry Lopez writes of a similar though considerably more ominous site in the far north: "South of Cape Bathurst and west of Horton River in the Northwest territories, bituminous shale fires that have been burning underground for hundreds of years make those coastal hills seem like a vast, smoldering heap of industrial slag."

Sheila described how the explosions trapped many workers underground. At the time I was too young to understand that all these men, even the survivors, were long dead and gone. Nor did I really understand the town name of New Castle—where was the *castle*? An active fantasy, then, for much of my early childhood: that the miners were somehow still down there, trapped in a cavern shaped like a castle, fending off the fire. My own version of hell, perhaps. In the years to come, the vision occupied my mind, feeding my childhood fears of the dark, of basements and crawl spaces. If I close my eyes, I can still see their figures etched on the walls of my memory—shadowy men in hard hats and soiled work clothes, their coal-smudged faces grim but determined to survive.

On that particular drive with Gabe and his family I managed to temporarily shed this new vision, turning my attention instead to our eager lookout for the wooden Smokey Bear sign at the entrance to the White River National Forest. When it came into view, us kids all cheered. Just as we always did when, several miles later, we spotted the metal roof of the A-frame cabin vaulted above the trees, silvering the sunlight. Arriving at the Mountain House meant free-roaming forty acres of horse pasture and aspen marsh, the forest and birds and our imaginations keeping us occupied. Up in that tranquil, creek-fed valley, we lived unconstrained by the boundaries and walls of everyday indoor life, the pale blue hatch of sky always wide open.

Wonders of the Deep

WHERE I GREW UP, WAY OUT ON THE LIMINAL EDGE of mountains and desert, the earth feels omnipresent. A larger-than-life main character. Whatever narrative you create about yourself here, she almost always makes herself known as the real protagonist.

Driving in through Glenwood Canyon from the east, geological strata pin-stripes the one-thousand-foot-high canyon walls, carved into visibility by the relentless force of the Colorado River. The Canyon's craggy sandstone and Cambrian rocks like precarious stacks of ancient history books authored by volcanoes and river-flow. Constant rockfall jackhammers the roads and long-span bridges on this section of Interstate 70—the most expensive stretch of highway ever built. The menacing roadside gullies all draped with chain-mail-of-the-gods fencing that may or may not shield cannon-ball-sized scree or forty-ton boulders from crushing your vehicle. As I write this, torrential mudslides—partly the result of deforestation by wildfires—have closed the entire canyon, indefinitely.

Just west of Glenwood Canyon, the eggy smell of sulfur registers high and heavy on your soft palate. On the right, you pass the Vapor Caves, a spa built in and above naturally heated, steamy caves, employed for various purposes beginning thousands of years ago by members of the Ute Indian Nation, long before white settlers invaded. Next, the Hot Springs Pool, my hometown's main attraction, commonly known as the largest geothermally heated pool in the country. Way above the pool, a couple thousand feet up the mountainside, you'll find the Fairy Caves—a maze of caverns dubbed the "Eighth Wonder of the World" by the first businessman to commercialize its marvels for tourists.

It's a desirable location, full of wonders—most of them sourced underground. We can guess this is what attracted the Utes—plus the obvious bounty of the Colorado River—then the miners who expelled the Utes, followed by the privileged classes, followed, in turn, by the working classes who catered to the wealthy or scratched out a living on ranches or in small businesses.

If you exit the highway, drive over the new bridge across the Colorado River, then make a sharp right-hand turn in downtown Glenwood toward the brick-hued foothills to the west, you may locate Red Mountain Drive. Take this road to its apex and you might spot a modest split-level ranch painted in drab earth tones. Now more than ever, tourists and deep-pocketed homebuyers flock to this valley. Even so, you'll observe how quiet and solitary life feels here, at the base of this parched mountainside. Crickets and the faint zing of traffic back on the highway. If you visited in the late 1970s or early '80s you might have seen a young boy, a latchkey kid with cowlicks and dimples, playing by himself in the yard, probably wishing he was thirty long miles away at the Mountain House, where his best friend lives what seems like a more adventurous life, surrounded by family. The latchkey kid's mother's nursing shift at the hospital doesn't end until well after 5:00 p.m. Before he moves to California, his father lives in a separate house across town, next door to an impulsive, lust-racked young man.

If you follow the latchkey kid on foot, he may walk up to the old, abandoned ski area behind his house, where he'll kneel down to examine bloodred stones or sheaves of mica mirroring the sun.

You might call the earth his companion, rocks his familiars.

With his thumbnail he scrolls through a chunk of mica's razor-thin, rectangular layers. He's awed by how easily he can pull them apart, like tearing pages from a billion-year-old text.

Sources

IN MY TEENS AND EARLY TWENTIES, I SPENT WEEKends exploring caves. Every Saturday, my body wedged in a different system—ribs, hips, and hands pressed against wet limestone. The earth exhaling cold rivers of air across my cheekbones. The loamy pong of mildew and mud. Water weeping through rock, plinking on sheaves of ice with symphonic rhythm.

I learned from guidebooks and older, more experienced cavers that you always wield three light sources against the absolute midnight:

- ~ Headlamp with heavy-duty, rechargeable batteries
- ~ Extra flashlight
- ~ Candles and waterproof matches in a ziplock bag

Light Source #1

GROWING UP IN THE 1980S, MY MOTHER AND I ATtended a small Episcopal church on a hillside near the hospital where she gave birth to me. The chapel's interior resembles the inverted hull of a wooden ship, an A-frame design with arched timbers. My mother graciously didn't make me go to Sunday school down in the drab basement. I stayed upstairs, where, during services, my attention drifted away from the hymnal and up to that elegant apex.

My favorite service was midnight mass on Christmas Eve, when new-fallen snow reflected the blue-white light of the moon above the chapel. Inside it smelled like frankincense and pine, and our minister always greeted us with his radiantly warm handshake. The most gentle man I'd ever encountered. I don't remember the particulars of his sermons. What stay with me are his kindness and the aesthetics of the room—recessed lighting and flickering candles bathing the vaulted pine ceiling in faint, numinous light.

Outburst #1

SIXTY MILES SOUTHEAST OF BURNING MOUNTAIN, along the same underground coal seam, lies Dutch Creek Mine No. 1. At 4:00 p.m. on April 15, 1981, an explosion ripped through a shaft in Dutch Creek. I was eight years old at the time, a latchkey kid. I remember letting myself into the house, turning on the TV to watch cartoons, only to find the emergency news report. A lantern of dread flickered inside my small rib cage, illuminating my distorted pictographic vision of the Burning Mountain disaster.

Three days earlier, combustion from two silo-sized rocket boosters launched the space shuttle *Columbia* on its maiden voyage. With the *Columbia* now orbiting two hundred miles above the earth, fifteen men were trapped half a mile below ground.

Dutch Creek Mine No. 1 was located above the town of Redstone, though operated mainly by workers from down-valley: Glenwood Springs, New Castle, Silt. Many were young men, just out of high school. The workings contained soft coal of a very high quality that was also exceedingly difficult to extract. Dutch Creek was widely known as a dangerous mine. The entrance was located at ten thousand feet above sea level.

After the accident, families of the Dutch Creek miners huddled around a campfire near the gate, waiting for word from rescue crews. One miner's wife spent a lonely night in the lobby of the nearby Redstone Inn. After thirty-three hours, the news was far from promising. "That big mountain ate my brother," said Vince Ayala, speaking of his younger brother, John.

Mine officials attributed the blast to an *outburst* of methane. One miner who witnessed the Dutch Creek blast and survived said it grew stronger and stronger, like a hurricane. Back in 1965, after a similar outburst killed nine miners, state officials named Dutch Creek as the most gaseous mine in the state, the second most gaseous in the country.

Light Source #2

DURING GRADE SCHOOL, I WAS PRIVILEGED TO SPEND winter weekends at Sunlight Ski Area—a mom-and-pop operation located on the site of a former coal-mining camp. Sunlight runs only two old-fashioned chairlifts, the green cable towers crowned with pulleys that broadcast a rubbery *squee-squee-squee* as you swing past. Back then my season pass cost seventy-five dollars, which wasn't exactly a bargain for my single mother, who worked two and sometimes three jobs to make ends meet after the divorce. Still, Sunlight was everything that the nearby resorts at Aspen or Vail were not—dads (mine included) skiing in jeans or snowmobile suits, parking lot full of pickups and Broncos and station wagons. No fancy hotels for miles. A place where townies and ranchers and middle-class folks from Denver could afford to ski.

Before I could walk, my mother rode the Sunlight chairlift with me bundled in her backpack. Once I got old enough, she glided down the slopes with me perched on her shoulders. Then, as I grew heavier, between her legs. Next, tied to a rope. Once she loosened the knot, she says I took off like a flash—always whipping through the trees or cruising the groomers in a full tuck, launching off every lip with my face pointed into the wind, doing my best to escape the earth's grip.

One New Year's Eve, friends invited my mother to ski in a nighttime torchlight parade. I watched from the lodge as she and forty or more skiers glided, single file, down one of my favorite runs, Storm King, each holding an emergency flare above their heads. Where effluent from a coal mine once poisoned these slopes, now a river of pink light ribboned down the mountain, illuminating the snow, reflecting off the pines.

Outburst #2

THE DAY AFTER THE 1981 DUTCH CREEK MINE DISASter, the latchkey kid walked from school to his father's house, across the street from Highland Park. The house being empty, he headed next door, to the home of two teenage brothers. The brothers and their parents lived in a single-level ranch house with a covered carport; their father delivered Hostess baked goods in a boxy white van, the mother worked as a receptionist. After school, the brothers introduced the latchkey kid to the music of AC/DC and Queen, the song "Cars" by Gary Numan. Over in the park, they all played a modified game of football called Smear the Queer. The brothers also knew where the former resident of the latchkey kid's house left a stash of old porno mags. The hiding place was in the attic above the garage, where the exposed pink insulation made the latchkey kid's inner elbows itch for a week.

According to current Colorado state guidelines, children under the age of twelve should not be left home, unsupervised. The latchkey kid was seven years old in 1981. In the nineteenth century, children as young as twelve worked ten-hour days in coal mines. In his essay "Sliver of Sky," Barry Lopez describes how certain men will exploit a divorce to find the vulnerable child within.

The older teenage brother, Clint, had a prominent Adam's apple. After school, Clint and the latchkey kid watched reruns of *Hogan's Heroes*, a TV sitcom set in a German prisoner-of-war camp during World War II. The actor Bob Crane starred as Colonel Hogan, a suave tactician who ran an underground resistance operation from an elaborate series of tunnels and bunkers below the camp. The mattress of his bottom bunk was actually a trapdoor to a secret staircase. Down in the lamplit bunker, they had control of the camp telephone switchboard and radio communication with allied forces. *Papa Bear calling Red Riding Hood.* Colonel Hogan often surveyed the camp with a submarine-style periscope. A doghouse and a tree stump served as escape hatches; the coffeepot doubled as a spy radio.

The day after the Dutch Creek Mine explosion, while their parents were all still at work, Clint and the latchkey kid sat on the couch together, watching *Hogan's Heroes*. During a scene when Colonel Hogan flirts with a blond German woman in a low-cut blouse, Clint moved closer, began running his fingers through the latchkey kid's hair. In 2002, *TV Guide* declared *Hogan's Heroes* the fifth worst TV show of all time, for the way it trivialized the suffering of the Holocaust and life in POW camps.

Later that week Clint played the song "Cars" by Gary Numan over and over, volume turned past eight, the chunky, robotic bass riffs shuddering the walls of his bedroom. Then Clint took the latchkey kid to see the remains of another porno mag spread in a messy pile behind the grocery store, the faces and bodies of the women wrinkled, torn, topographic with rain damage. Bob Crane, the star of *Hogan's Heroes*, was both a connoisseur and a creator of homemade pornography. As portrayed in the 2002 biopic film *Auto Focus*, Crane may have kept his stash hidden in the basement.

Colonel Hogan and his men carved so many tunnels and bunkers, it's surprising the entire POW camp didn't cave in on itself. When a visible hole sometimes did appear, the Heroes convinced Colonel Klink they were hot springs or natural caverns.

At school, the latchkey kid drew obsessive pictures of tanks with knobby tires and parallelogram-shaped chassis, perfect for carving into the earth. The tanks excavated a mazelike system of tunnels that were otherwise devoid of human life. Later, at the basketball court, he played an obsessive game in his head: *make the next shot and you might escape this burning mountain, make the next shot and you might survive.*

Light Source #3

THE SUMMER AFTER MY FRESHMAN YEAR IN COLLEGE, I worked construction and spent most weekends up at the Mountain House with Gabe. One Saturday in July we decided to meet instead above the town of Eagle at the trailhead to Fulford Cave. A few miles up the trail, we located the metal culvert the Forest Service had angled into the ground and threaded with a ladder, to keep novice cavers like us away from the other, more dangerous vertical entry shaft.

After hiking through the hot afternoon, the backs of our T-shirts felt wet against our daypacks. When we flipped open the plexiglass lid to the culvert, we stood in awe for a moment of Fulford's humid, musty breath. Against my sun-warmed hands and face it registered at least twenty degrees cooler than the outside air. We steeled ourselves and packed our bodies into the slanted tube that tapped this unseen world with its own weather system.

Through a series of horizontal chambers and passageways, we reached a subterranean creek that shimmered, translucent and clean, beneath our Petzl headlamps. Later, after stemming up a vertical chimney, we nearly dead-ended against a granite wall with a two-foot-wide crack running horizontally up its sheer face. We chocked our torsos into the crack and inched up to the Cathedral Room—a grand chamber, intimidatingly large, where stalactites met stalagmites to form what looked like a pillowy, limestone pipe organ.

When, two hours later, I heaved open the culvert lid, the torrent of sunlight overwhelmed my senses. My pupils struggled to adjust as Fulford delivered us back into a new and radiant world, where afternoon sun on quaking aspen leaves glimmered brighter than Christmas trees, so much that we had to shield our eyes with muddy hands. The air felt oven-warmed and parched compared to the cave's damp chill. Our clothes dried as we rambled down the trail; the sensory disorientation morphed into a subtle feeling of liberation as we slowly took in the sky's capacious blue vault.

The only experience I can liken it to, roughly, was visiting Barcelona's Sagrada Familia cathedral two decades later with the woman who became my wife.

We arrived before dawn, waiting for opening time in the near-dark. Among the first to enter, we wandered through molten pools of greens, golds, and reds—the effect of Spanish morning light projecting stained glasswork onto the stone floor. The one-hundred-and-fifty-foot ceilings kaleidoscopic with light and color. Most captivating to me was the brilliant, orange-tiled portal high above the altar, like the burnished interior of a cockle seashell. Capped with a clear circular skylight, the portal focalized the Mediterranean sunlight and my own attention. Just above the portal's mouth, a cluster of golden tiles—constellated into the pattern of a geometric tree or a narrow pyramid—pointed up into the skylight like an indicating arrow for the divine.

I stood beneath it, rapt and motionless, until the cathedral overflowed with other tourists. I hardly noticed the river of people eddying around me because the portal had overtaken my earthbound soul, and like an inverted whirlpool, coaxed it skyward.

Outburst #3

DAY AFTER DAY, CLINT RAN HIS FINGERS THROUGH the latchkey kid's hair. Until he stood up and closed the curtains, unbuttoned his jeans. And then they were like the characters on television—if anyone approached the door to their barracks, Clint quickly pulled their clothes back on, flipped the couch cushions over to hide the secret crooked staircase that led down into the bowels of his covert operation. This is where the latchkey kid lived for the next decade or more, bunkered deep in the cave of his body, knowing that if he was found out, it would be the end, a kind of annihilation, *Red Riding Hood calling Papa Bear.*

The verb *exploit* arose circa 1400 in prototypical forms *espleiten* and *esploiten*, meaning "to accomplish, achieve, fulfill," from the Old French *esploitier* and *espleiter*, which themselves derived from *esploit*. The sense of "use selfishly" was first recorded in 1838, from French, likely extended from use of the word with reference to *mines* and *mining*.

The actor Bob Crane was murdered in a hotel room in Scottsdale, Arizona. He was in town with his touring dinner-theater production entitled *Beginner's Luck*. Crane's case remains unsolved, though suspicion falls heavily on John Carpenter, an electronics expert who filmed and participated in dozens of Crane's sexual exploits. According to police records, Crane was bludgeoned with a heavy object, possibly a camera tripod.

Saint Augustine writes, "We make ourselves a ladder out of our vices if we trample the vices themselves underfoot."

In 1978, John Carpenter was acquitted of the murder of Bob Crane, though police found spots of dried blood in his car, and a waitress who served Crane and Carpenter lunch on the day of the murder described the mood at the table as tense. Carpenter was retried and again acquitted in 1992; he professed his innocence until his death in 1998.

During a civil case against Mid-Continent Resources, a total of 419 safety violations were found to have occurred at the Dutch Creek Mine between 1976 and 1978. According to miners, being in a mine shaft during a methane explosion is the equivalent of being inside an exploding gun barrel.

New York Times critic A.O. Scott wrote that the Bob Crane biopic film *Auto Focus* "gets to you like a low-grade fever, a malaise with no known antidote . . . There is plenty of nudity in *Auto Focus*, but you can always glimpse the abyss behind the undulating bodies, and the director leads you from easy titillation to suffocating dread . . ."

In 1984, when his father was out of town and the teenage neighbors had come over to "babysit," the latchkey kid awoke to find Clint on top of him. They had all fallen asleep in the living room in front of the television. The latchkey kid was several years older now, almost eleven. He kicked Clint away and retreated to his bedroom. The next morning, while Clint was in the bathroom, the latchkey kid considered telling Clint's younger brother, Fred. Informing. *Betraying the operation.* The desire to spill the truth was like a buildup of noxious gasses in his head and chest, but he could not press the shame through the valve of his throat. After the 1981 explosion, a cap was placed permanently over the Dutch Creek Mine entrance. This was when the latchkey kid learned the final lesson of silence—the avoidance of any outburst that might reveal the underground operation in which he was trapped, *that big mountain ate my brother.*

In one episode of *Hogan's Heroes*, a German mole tried to penetrate Hogan's subsurface resistance movement. Hogan and his men blindfolded the spy, feeding him false information and leading him to believe that their operation was under the water tower—the one building in the entire camp under which they had not excavated.

The metaphoric use of the word *mole* as "one who works in darkness" is from circa 1600. The term's "spy" connotation was first recorded in 1974 in the novels of John Le Carré, but had likely been used as such since the early twentieth century.

There's another word, similar to *mole*, that derives from Old English *mal*, meaning "to spot, mark, or blemish." Meaning "to stain or defile." The late-fourteenth-century usage refers specifically to dark marks on the human skin. The word also means "to bother, interfere, or annoy." *To make indecent sexual advances toward.* Possibly related to the more animalistic connotation of *mole*, as in: burrowing, darkness.

As in teeth and claws.

As in something you coax up from underground, its eyes white and atrophied from so many years of tunneling.

Buried Memories

IN DECEMBER 1945, WHILE DIGGING FOR FERTILIZER inside the Jabal al-Ṭārif caves near the town of Nag Hammadi, an Egyptian farmer named Muhammed al-Samman discovered twelve leather-bound papyrus codices sealed inside an earthenware jar. The works became known as the Gnostic Gospels; they'd likely been buried for protection from Bishop Athanasius, who ordered the burning of noncanonical texts after the creation of the strict Nicene Creed in A.D. 325. Line 70 of one Gnostic Gospel quotes Jesus as saying, "If you bring forth what is within you, what you have will save you. If you do not bring forth what is within you, what you do not bring forth will destroy you."

Exhibit

INSOMNIA AFFLICTED ME THROUGHOUT GRADE school, nighttime like a snare. After my father remarried a woman whose teenage son had just died in a car accident, I moved into his former bedroom, where I kicked at the sheets all night, my boyheart crashing against the boxspring mattress. During the school day I was either hyperactive or half asleep at my desk, and constantly in trouble with my teacher, Mr. Prescott. During a parent-teacher conference, Mr. Prescott informed my parents that I was underperforming academically and exhibiting problematic behavior.

I can't recall a single thing I learned or read that year.

The only thing I remember is a trip to the Glenwood Springs Frontier Historical Museum, located in an old Victorian house a few blocks from the elementary school. We climbed the creaky wooden staircase to the second floor, where we entered a replica nineteenth century mine shaft: black curtains, construction paper, plaster-of-Paris boulders arranged around two antique display cases. Fake orange lamplight flickered within the dark walls. Inside the glass displays sat carbide lanterns, like miniature satellite dishes attached to old hard hats. Various mining implements: shovels, picks, scales, a bundle of faded-red dynamite. On the wall: sepia-toned photographs of miners. Dark hair and mustaches that contrasted with their phantom-white skin and sunken cheeks. The men and their mules looked bone-tired, weary.

In the background, tape-recorded sounds played on a loop—the sharp clank of pickaxes on stone.
    Rockfall.
    Distant explosions, men's voices.
    Echoes.

Sliver

IN HIS *Harper's Magazine* ESSAY "SLIVER OF SKY: Confronting the Trauma of Sexual Abuse," Barry Lopez describes the lingering effects of sexual assault on survivors. "I felt more or less continually sea sick," he writes, "confronting every day a harrowing absence within myself. I imagined it as a mine shaft of bleak, empty space, which neither the love of a spouse nor the companionship of friends nor professional success could efface."

In February 2015, after a weekend writing workshop in Newport, Oregon, Lopez invites me to lunch. Later that evening he'll have a beer or two and regale his wife—the memoirist Debra Gwartney—and me with more light-hearted stories, like the Polynesian navigators he witnessed dipping their testicles into the ocean to gauge its salinity. But over our lunch, the topic is writing, and the tone remains more serious. When he asks what I'm working on, I share how his "Sliver of Sky" essay gave me the courage to confront some of my own deeply personal material. I tell him about buried memories, a subterranean theme.

"Most of it revolves around mining," I say.

"That's essentially the origin of it all," he responds. Meaning industrialization, the doctrine of manifest destiny, the subjugation and devastation of Indigenous cultures and peoples, our entire Anthropocene era.

Origins

THE SOUTHERN ROCKIES WERE ORIGINALLY NAMED the Shining Mountains. (Upon first learning this more artful and evocative name, I pledged to use it in place of the more banal *Rocky Mountains*.) The original inhabitants of what is now Colorado gave the range this older, better name. They call themselves *Núuchi-u*, or "the people."

The Núuchi-u are now more commonly known as the Utes. Before colonization, they inhabited most of what is now Colorado, including land around the geothermally heated Vapor Caves and the Hot Springs Pool. White settlers eventually wrested control of the caves and the springs, turning them into tourist attractions, but the Utes consider them places of healing and renewal.

In 1849, the Ute Nation of central Colorado signed a treaty with the United States. In return for accepting the U.S. government's authority, the Utes began receiving $5,000 a year, and were promised all their lands—comprising most of Colorado's interior with all its Shining Mountains and vapor caves and hot springs and hunting grounds—for as long as "the rivers should run and the grass grow."

From *Glenwood Springs: The History of a Rocky Mountain Resort*: "It cannot be said that they [the Utes] 'owned' the territory in which they resided. Their beliefs held no concept of the ownership of land." The same might be said for their concept of ownership of what lies beneath the land, as in minerals, precious metals, vapor caves.

In the late 1850s miners and prospectors rushed the Colorado territory by the thousands after gold and other mineral wealth was discovered in what we no longer call the Shining Mountains. Various U.S. military outposts were created to control and protect Utes and their land. By 1879, the region was the scene of manic prospecting and mining exploits; as many as five thousand migrant miners surged in on a daily basis. At this time, Utes still held the area around what is now Glenwood Springs, where they continued using the vapor caves and hot springs, even amid such frantic excavation.

From an editorial in the *Denver Times*, circa 1879: "Either the Utes or we must go, and we are not going. Humanitarianism in an idea; Western Empire is an inexorable fact."

Pliny the Elder, who, despite his own obsessive fascination with gemstones, wrote of the practice of mining precious metals and stones from the earth, "The things that she has concealed and hidden underground, those that do not quickly come to birth, are the things that destroy us and drive us to the depths below . . ."

Chief Ouray, leader of the Utes during the gold rush, was later quoted as saying, "Agreements the Indian makes with the government are like the agreement a buffalo makes with a hunter after it has been pierced with many arrows. All it can do is lie down and give in."

My hometown was originally named Fort Defiance by white settlers and prospectors, as a means for publicly registering their grievances with the Utes who refused to leave, and with the U.S. government that (only too briefly) codified land rights for the Utes. The name was later changed to the more anodyne Glenwood Springs, though much of the town was built on mining wealth.

In *Meet Joe Copper: Masculinity and Race on Montana's WWII Home Front*, author Matthew L. Basso describes the role of virulent masculinity in shaping or "taming" the West, and how the mining industry was a key force in forging our cultural conception of "real men."

Growing up, one of my few sources of information about Ute culture was via a community theater production at the local high school, with friends of my parents in redface, speaking in crude dialect. Up on stage, they lounged on deck chairs, some in fake, feathered war bonnets, joking about their yearly "tourist vacation" to the Hot Springs Pool. Bursts of laughter detonated out among the audience.

The nearby town of Meeker, Colorado, was named after one Nathan C. Meeker, a former *New York Tribune* journalist who tried and failed to organize a utopian society—a "Union Colony of Colorado"—based on high moral principle and a strict code of sobriety, with backing from east coast editor Horace Greeley. When Meeker's scheme failed (due to his abrasive, holier-than-thou demeanor), he went nearly bankrupt. To secure his future, he got himself appointed as an Indian agent at the White River Agency, which provided "aid" for regional tribes. Meeker proved equally unpopular with the Utes. He wanted to subjugate them with agriculture—to beat their weapons into plowshares.

Meeker eventually ordered the destruction of the Ute's valued racetrack and the killing of their horses. A Ute medicine man named Johnson subsequently shoved Meeker through a doorway and onto a hitching post. In response to this single shove that was the result of decades of shoving, of being shoved, Meeker called the governor. The governor summoned federal troops, now in opposition to the Utes.

Before the army arrived, Utes attacked the White River Agency encampment. They killed Meeker, who might rightly be considered the low-grade Colonel George Custer of Colorado. When Chief Ouray heard about the skirmish, he was reported to have wept openly, considered suicide.

After what white settlers deemed the "Meeker Massacre," government troops forcibly removed Ute Nation members from the mineral-rich state of Colorado. Many were coerced onto reservation lands in eastern Utah, where they have less abundant access to hunting and water, but where they continue to adapt, as they always have. After their expulsion, the vapor caves were increasingly utilized by the white settlers. The (unimaginative) new name: Old Cave No. 1.

All it can do is lie down.

Time and again, gold rushes and mining industry catalyzed the breaking of treaties and forced removal of Indigenous Americans from their ancestral lands. White settlers perpetrated a similar expulsion upon the Nez Perce in eastern Oregon. In the wake of a gold rush, U.S. government–backed forces chased Chief Joseph and his people off their homelands.

At a graduation ceremony at the university where I teach, I told a Native American colleague how my research seemed to indicate that the Ute Nation and others had somewhat favorable treaties up until the Colorado gold rush.
    Her facial expression turned severe.
    "There is no such thing as a *favorable treaty*," she said.

The American Indian Wars started in 1609 and "ended" some 315 years later. In popular media, the twenty-year war in Afghanistan is generally cited as America's longest military conflict.

From Genesis 1:26: "Then God said, 'Be fruitful and multiply, and fill the earth and subdue it; and have dominion over the fish of the sea and over the birds of the air and over every living thing that moves upon the earth.'"

Crystal Cavern

AFTER MY PARENTS' DIVORCE IN THE EARLY '80S, MY mother spent a year living in Denver, where she earned her master's degree in nursing. She rented a room from an amiable schoolteacher—the first fully out lesbian I ever encountered, who showed me that queerness does not equal ridicule or exile, as it seemed to back in Glenwood Springs. I felt safe when visiting my mother and her in the city, and on the weekends when my mom made the three-hour drive home. But I spent most of that year at my father's house; he often fell into sour moods during this era of his single parenthood. Alcohol sometimes exacerbated his gruffness. Other times my own misbehavior and hypersensitivity tested his patience. Then there was his neighbor's house where the teenage brothers lived, always looming. It never even occurred to me to tell my mother what went on over there, though her vocation involved protecting the vulnerable.

One bright, cold Saturday, she drove me to the Denver Museum of Nature and Science, to visit the Gems and Minerals exhibit.

"Colorado was founded on mining," a placard informed us as we entered the exhibit.

I remember Tom's Baby: an eight-pound nugget of crystallized gold unearthed near Breckenridge in 1887.

In a darkened room, fluorescent mineral specimens transformed into exotic shades of neon green, purple, and red under a push-button-operated ultraviolet light bulb.

Nearby, encased in a large cubby hole behind thick glass, an installation called The Grotto contained a fully preserved cavern of crystals imported from Mexico's Naica cave.

The interior was an iridescent, billowy cloudscape—a confectionary wonderland of stalactites and stalagmites like puffs of spun sugar or tufts of glittery wool.

It seemed to go on forever, the cavern growing dark as it curved to the right, creating the illusion of infinite depth. I wanted to crawl inside and curl up in the back corner, make my home inside its soft, scintillating world.

I imagined I could live safely in that splendor. The crystals emanating a protective force field, thick glass preventing any unwanted hands from reaching inside.

Geologist's Field Hardness Scale

Hardness	Common Tests	Mineral
1	Easily scratched by fingernail	Talc
2	Scratched by fingernail, but not easily	Gypsum
3	Scratched by penny	Calcite
4	Scratched easily by knife (4.5), but will not scratch glass	Fluorite
5	Difficult to scratch with a knife (5.5); barely scratches glass	Apatite
6	Scratched by steel file	Orthoclase
7	Scratches steel and glass	Quartz
8	Scratches quartz	Topaz
9	Scratches all minerals except diamonds	Corundum
10	Can only be scratched by another diamond	Diamond

(1) In a different kind of story, Clint might have dug his fingernails into my small wrist, warned me never to tell anyone. He didn't have to, so intense was my shame. I didn't betray the operation for almost ten years, when the truth came spilling out in a therapist's office one thousand miles west of Colorado.

(2) As a kid I pressed my own thumbnail into the thin skin of my wrist until I made a bruise or drew blood. I did it

just to feel something. Or to ward off the dread that might creep up under my comforter during the night.

(4.5) When I was thirteen I etched a bloody anarchy symbol into my forearm with a pen knife, music by T.S.O.L. and Agent Orange crashing out of my cassette player. It burned and throbbed but getting at the ache buried inside me—earning a big circled A on my own pain test—felt good somehow. I savored the sting, like I would later in my twenties, when a tattoo artist inked the image of an antique fountain pen on my inner bicep.

(5.5) One night in my forties, in a fog of depression and having just learned my mother had cancer, I tried cutting the inside of my bicep with a kitchen knife. Hurting myself was harder than I thought, and I didn't want to ruin my fountain pen tattoo. What emerged was barely a scratch.

(7) For a period in the 1990s, instead of feeling scared all the time—never knowing when dysregulated panic might knife through me—I tried to be hard. Diamonds were never my thing, but quartz-level imperviousness seemed achievable. I climbed two-hundred-foot granite faces and hammered on my Ludwig drum kit and skated Burnside until I bled all over my Dickies and Spitfire T-shirts. Listened to Drive Like Jehu and Fugazi while I drove alone around the city, inwardly fuming, punching my knuckles raw on the steering wheel. It was about driving something down, trying to scratch at something instead of feeling scratched from the inside—grinding my skateboard axles on steel coping felt better than being ground down by fear and depression. To build skateboard ramps, I learned to cut sheet

metal with a carbide saw blade that spewed a fountain of sparks. My hands and elbows forever inscribed with burn marks, road rash, scratches, and scars—no matter how I tried to fake it, my skin's a text that always reads low on the hardness scale.

Fusion

QUARTZ IS A FORM OF SILICA—A COMMON ELEMENT that makes up around 60 percent of the Earth's crust.

Silica is known to exist throughout the universe and in meteorites older than our solar system. It's abundant on sandy beaches, within the earth's core, inside our cell phones and computers.

Silica was very likely formed by supernovas—stars much bigger than our Sun that imploded after depleting the fuel that burns in their core. The collapse creates a nuclear explosion of such colossal scale that it can fuse atoms together to birth heavy elements like copper, calcium, and silica.

In essence, supernovas become a kind of cosmic forge that spews elemental dust across the universe. Over millennia this dust forms new planets and stars.

When we hold a quartz crystal in our palm, feel its faint pulse against our skin, when we peer out a glass window or walk barefoot on a sandy beach or text *I miss you* to a loved one—in each of these instances, we engage with elements created by exploding stars that blazed billions of years ago.

Dead Work

IN THE EARLY TWENTIETH CENTURY, COLORADO MINers were compensated per tonnage of coal they produced, while laborious and essential tasks like reinforcing unstable mine walls and ceilings with timber, also known as "dead work," were almost always unpaid. Imagine, if you will, strong-arming modern-day office workers into installing fire escapes and seismic retrofits on evenings and weekends. Without compensation for safety work and with minimal governmental oversight, the death rate in Colorado mines was much higher than average. Cave-ins, rockfall, explosions, and down-shaft falls were all daily risks for miners.

It wasn't just men who put themselves in harm's way. "Breaker boys," age twelve or younger, worked long hours sorting coal chunks, with a company-employed master at their backs, enforcing productivity. Breaker room atmosphere was so thick with coal dust the boys could barely see; many developed black lung by young adulthood. Slightly better off were the "nippers," who sat all day in the bowels of a coal mine, waiting to open the door for approaching coal cars. Often these boys were forced into labor to support families after their fathers sustained severe injuries or died in the mines. This is generally not something mentioned in displays of unearthed minerals and gems at places like the Denver Museum of Nature and Science.

John D. Rockefeller Jr. co-owned Colorado Fuel & Iron Company (CF&I), one of the most powerful operations in the state. The Rockefeller family also owned much of the infrastructure in the Newtown Creek facilities, in North Brooklyn, from which seeped the largest and perhaps least-remembered oil spill in American history. To extract his coal, John D. Rockefeller Jr. imported workers from all over eastern Europe, Africa, Mexico. He specifically recruited men who spoke different languages, their inability to communicate preventing them from organizing to demand better conditions.

Compared to the ramshackle life of solo prospecting, mining corporations did improve living standards and health care for many miners and their families. But the conditions were very often prison-like. Only company-approved literature and films were permitted. Armed guards and barbed-wire fences were employed against theft, but more often served to prevent miners from getting out of line or shopping anywhere but the company store, where distant owners like Rockefeller severely gouged prices.

Any dissent or breaking of rules often resulted in eviction from the camp. In *Angle of Repose* by Wallace Stegner, a loyal Cornish miner loses his job, and a company-hired arsonist burns his house down, all for the offense of purchasing some rope from an outside store. It's a fictional account of very real practices, and a salient example of how my Cornish ancestry includes both the exploiters and exploited.

The harsh conditions and unjust labor practices inevitably led to civil unrest. Activist and writer Mother Jones paraded the streets of Trinidad, Colorado, arm in arm with miners' wives, demanding rights. Miners began organizing for safer working conditions, collective bargaining rights, 10 percent wage increases, shorter workdays, payment for "dead work," the right to use any store, school, or doctor, and an end to companies' carceral security details.

When major coal companies refused these demands, workers organized the Colorado Coalfield Strike of 1913–1914. Rockefeller and other mine owners promptly evicted any dissenting employees—*they drove us miners out of doors*, Woody Guthrie later sang in a ballad composed to honor the Colorado miners. The United Mine Workers helped organize a temporary tent city near Ludlow, to house displaced strikers and their families.

Rockefeller hired militiamen to protect scabs and harass striking miners. They regularly fired at random into the camp, wounding innocents and inflaming terror. Miners and their families dug pits beneath the tents for protection—*carried our young ones and pregnant women / down inside the cave to sleep*.

The governor called in the Colorado National Guard, who largely sided with mine owners to help dismantle the protest. An armored railroad car mounted with a machine gun, nicknamed the "Death Special," patrolled the camp's perimeters. After a standoff lasting several weeks, militia members shot the miners' leader, Louis Tikas, in the back. They left his body in plain sight of passing trains, as a warning.

On April 20, 1914, after Tikas's murder, a major firefight broke out and raged until evening. By nightfall, the anti-union militias attacked the camp and began looting and pillaging—*soaked our tents with your kerosine.* Four women and eleven children hid in a pit beneath one tent. After militiamen torched the canvas, two of the women and all eleven children suffocated.

I made a run for the children but the fire wall stopped me.

In response, miners sent out a statewide call to arms. A ten-day guerilla war ensued at mines across the region. Now known as the Colorado Coalfield War, it was among the deadliest battles between American citizens since the Civil War. Seventy-five people were killed; the fighting ended only when President Woodrow Wilson sent in federal troops.

In hearings conducted by the United States Commission on Industrial Relations, John D. Rockefeller Jr. testified that, even after learning of the atrocities, he "would have taken no action" to prevent his militias from murdering miners and their families. He wasn't held criminally accountable for the massacre, and to rehabilitate his subsequent reputation as one of the most hated men in America, he hired a public relations firm to spread various falsehoods about the striking miners and to cement his legacy as a magnanimous philanthropist.

The Commission's report suggested reforms and supported bills that eventually established what we now assume as given for most workers: eight-hour workdays, weekends, overtime pay, a ban on oppressive child labor.

Quarry

THEY WERE LIKE CALVED ICEBERGS, OR A RANSACKED library of giant white books. The marble slabs' edges striated from the quarry drills.

The deepest place I'd ever seen. The marble quarry like an empty socket in the earth, the bottom filled with milky, aquamarine water. As I peered over the edge, my grip tightened on my father's hand. This was during the time in my childhood when I was terrified of basements and dark places, when I had trouble sleeping through the night.

> Marble is recrystallized limestone, normally white... Limestones and dolomitic limestones may be slightly altered by percolating waters and are often called marbles, but true marbles are the result of metamorphism involving heat and pressure. —FROM *ROCKS AND MINERALS, A GOLDEN NATURE GUIDE*

I don't remember if it was actually sunny that afternoon but given Colorado's three hundred average days of sunlight per annum, we can assume shafts of auric light slanted into the quarry.

My father explained how quarrymen sliced off white slabs of marble, lassoed them with cables, winched them up from the pit. Marble, he told me, is Colorado's official state rock. He also explained that marble from this quarry was of exceptional quality and was used to sculpt structural elements of the Lincoln Memorial in Washington, D.C.

Melvillean scholar Lewis Mumford was skeptical of the Lincoln Memorial's colossal proportions, its "classical serenity." "In truth," Mumford writes, "the Lincoln memorial was a particularly clever piece of imperialist propaganda." In Christopher A. Thomas's *The Lincoln Memorial and American Life*, he describes the temple as "a confection of a cultural and political elite bent on stripping Lincoln of his earthly imperfections." Imperfections, we can assume, such as Lincoln's dim view of race relations and his ongoing, genocidal war against Native Americans. His regular descents into gloom and depression. And perhaps his notorious Adam's apple, like something trying to push its way out of his throat.

On the drive home from the marble quarry, my father and I stopped at a house up Divide Creek, near Colorado Mountain College. A beautiful home with a swimming pool, large deck, picture-frame windows looking out over the scrub oaks and sagebrush. The house belonged to a friend of my father's, a blind man from France named Jules Amis. He was in my father's Rotary Club; Dad said he knew everyone by the sound of their voice.

My father described Jules Amis as "a miraculous guy." He played golf purely by sound, with helpers standing on the green, ringing a bell. He once challenged my father to eighteen holes. When my father accepted, Amis said, "Very well, then. Meet me on the course at midnight." He helped launch one of the original blind skier organizations, called BOLD: Blind Outdoor Leisure Development. He was a sculptor and engraver who may or may not have engraved the eagle on a piece of ubiquitous American currency. He could fix anything, including the broken pedal of a stationary bike that I tried to ride on his outdoor patio.

During our visit we watched Jules work on a life-size sculpture of a human figure in his studio off the garage. He sculpted with marble mined from the same quarry my father and I had just visited. He carved and shaped with power tools, his face coated in fine white dust. He used bare hands during the finishing work, moving them up and down the marble face, slowly rousing the human figures in the stone up from their long sleep.

Detractors of the Lincoln Memorial say it's too polished, too naive in its nationalistic, hypermasculine hero-worship. Perhaps in a related but inverse vein as literary critics and activists who warn against the extraction of facile "lessons of inspiration" from those with disabled bodies.

I hope what I'm doing is less about memorializing or drawing inspiration from Jules Amis, and more about the lesson of art as something born from heat and pressure and darkness. Rebecca Solnit, from *The Faraway Nearby*: "Darkness is generative, and generation, biological and artistic both, requires this amorous engagement with the unknown, this entry into the realm where you do not know quite what you are doing and what will happen next."

The truth was that something about Jules Amis scared me. The unpredictable way he ambled and stomped across the wooden deck. His French accent and thunderous voice, always echolocating. How he rummaged around in his toolbox, found a wrench to fix the bicycle pedal, eyes on the tips of his fingers. All those marble figures ghosting around the house. I didn't understand his sculptures, though I wanted to be part of them, somehow. Perhaps my first exposure to the transformative power of art, employing raw material from the yawning sub-terra.

Historically, mining has always fueled both creativity and destruction. In *Bluets* author Maggie Nelson references *Sar-e-Sang*—"the Place of the Stone," a mine in Afghanistan where workers unearthed lapis lazuli beginning in the seventh millennium BC. Ancient Mesopotamians and Egyptians used the brilliant blue stones in King Tut's funeral mask; Cleopatra may have ground it into powder for eye shadow. Much later it was pulverized into the most expensive of all pigments: ultramarine. Traces of ultramarine directly from Sar-e-Sang exist in the paintings of Titian, Vermeer. Mary's blue robes in most Renaissance paintings would not emit their effulgent brilliance without the act of mining.

I can see both sides of the Lincoln Memorial debate. For me, what makes Lincoln memorable was his psychological fortitude, how he carried on despite the misery in his past, his tendency toward what Melville called "the hypos." Maybe, then, all those cut and quarried slabs—*true marbles are the result of metamorphism*—winched up from such depths, are a fitting tribute, even if the statue itself exaggerates his physical proportions, downplays that outrageous Adam's apple.

The word *sincere* translates roughly as "without wax," dating back to Roman times, when some marble salesmen covered up blemishes in their products with cosmetic wax. Other, more trustworthy merchants presented the stone as it was, without masking the flaws.

My father had his own faults, his lack of attention to certain aspects of child-rearing or road safety, but he was the most devoted and fun-loving parent I could ask for, always taking me to quarries, abandoned mines, the High Country Gems and Minerals shop. A week after that first visit to the marble quarry, he put up with me when, during his afternoon nap, I snuck into his bedroom and set the digital alarm clock thirty minutes forward, hoping he'd wake up and take me on one of our excursions. Perhaps this is where I learned to engineer time and chronology and memory for my own purposes.

I never took to sculpting, but years later, on the advice of a therapist, I began obsessively grinding words into marble-patterned composition books. It was there that, from a slab of buried memories, I carved out the awkward, lanky shape of our teenage neighbor, with his prominent Adam's apple that reminded me of Abe Lincoln's. And how, after school, he coerced me down into a different kind of quarry—the mouth of the cave that led to my childhood insomnia and fear of the dark, my lifelong headlamping through the hypos.

Though it frightened me a little, Jules Amis's accent also reminded me of Corporal LeBeau, the French resistance operative from *Hogan's Heroes*—the TV show always on when Clint forced me into underground labor. LeBeau was claustrophobic, but also the smallest of Hogan's men, so he was often enlisted to hide in tiny, enclosed spaces: the dumbwaiter, the safe in Klink's office, various tight shafts and crawl spaces.

He was often alone in the dark, sacrificing for the resistance.

> Metamorphic rocks are rocks which have been changed. The changes usually bring about a new crystalline structure, the formation of new minerals, and sometimes a coarsening of texture.
> —FROM *ROCKS AND MINERALS*,
> A GOLDEN NATURE GUIDE

It rained on that drive after I woke my father up from his nap. He whistled as always while driving, steering with his knee over gravel and brick-red mud puddles. I don't remember where we went, exactly, but I want to believe I asked him if we could visit Jules Amis again, to watch quietly as he gifted quarried marble with imperfect human form.

Things That People Who Shine Can See

BEFORE PRODUCTION BEGAN ON THE CLASSIC '80S horror film *The Shining*, filmmaker Stanley Kubrick hired a team to research the gold rush–era history of the western United States and the subsequent forced removal and murder of Native Americans.

In the film, Jack Torrance, famously played by Jack Nicholson, takes a job as the winter caretaker for the remote Overlook Hotel—where he hopes the extreme isolation will help rekindle his writing career—with his wife Wendy and son Danny in tow.

During an orientation, the polished Overlook manager tells Jack Torrance that construction workers had to fend off "Indian attacks" as they erected the hotel directly above a Native American burial site.

As many critics theorize, the river of blood that floods from an Overlook Hotel elevator in a notorious scene from *The Shining* can be read as the Indigenous blood on which the nation was built.

The blood that white settlers concealed down in the desecrated burial site that is forgetfulness and repression of historical sins—a collective *overlook*.

Original European posters for *The Shining* read, "The Wave of Terror That Swept Across America."

I first watched *The Shining* at my friend Sean's house, when I was maybe nine or ten.

In Sean's basement, on a prototypical large-screen television, I witnessed those ghostly twin girls in blue dresses, their hacked-up bodies, a psychotic Jack Nicholson chasing his family through the Overlook's empty halls.

Sean and his older sister seemed unaffected, but the horror I experienced felt infinite, intractable.

Around 1:00 in the morning I called my father to come pick me up.

That night and every night for months afterward I wandered the hallways of the dim, uninhabited hotel that is dread-induced insomnia, small heart axing against my ribs.

In Stephen King's original novel version of *The Shining*, hotel chef Dick Hallorann was sexually abused by his grandfather when he was five.

This primal wound sparked Halloran's ability to "shine," meaning he could psychically communicate with others like him without speaking a word.

As Halloran explains to Danny Torrance, all the bad things that happened in the Overlook over the years left ghostly traces that not everyone can see, but "that people who shine can see."

Jack Torrance once yanked Danny by the arm so hard it dislocated his shoulder.

Some critics suggest Jack may have also sexually abused Danny, which would further explain Danny's own ability to "shine."

The Shining, then: a narrative about survivors' abilities to bear witness to historical injustices and the suffering of those who survived (and those who didn't), long before.

And about the stealthy yet profound communication possible among living survivors.

When we played alone in my father's house after school, Sean and I sometimes took off our clothes and pressed our bodies together.

My first experience with something like mutually consensual sex, though I think I wanted it more than he did.

One day on his front steps, Sean told me that his parents said it would make him sick to keep doing what we were doing.

He closed the door in my face while I sobbed.

Later, as a teenager in therapy, I had a hard time teasing apart my actual queerness—my first consensual partner was another boy—from the assaults perpetrated upon me.

Sean and I were just kids, but for me there was something beyond the childhood exploration and possible replication of Clint's abuse.

In my teens I lived in East County, San Diego, where I feared the ridicule and violence I would have faced had I consciously expressed anything but rigid heterosexuality. Upper classmen in my high school bragged about late-night drives to Hillcrest to beat up "fags."

I live in Oregon now, an easy hour drive from the Timberline Lodge, where most exterior shots of the fictional Overlook Hotel were filmed.

All my real-life associations with Timberline are joyful—snowboarding and hiking at the base of cloud-haloed Mount Hood.

Nothing about the place frightens me.

And I now find it oddly comforting to watch horror films with their self-contained architecture of terror, in which I can temporarily outmaneuver my real-life anxieties.

Nor does my queerness trouble me, having let go the struggle to fit myself into an either-or category when it comes to sexuality.

Does anyone who watched *The Shining* at a tender age, though, ever fully escape the looping, labyrinthine hallways of the Overlook Hotel?

One particularly unhappy winter in my mid-forties (the same winter the forty-fifth U.S. president was first elected), I posted a video clip of *The Shining* elevator spilling its thousand gallons of blood.

It was my ironic version of a personal "year-in-review" highlight reel, so popular on social media around the winter holidays.

Clearly the post was in poor taste, but *I* thought it was hilarious.

It received thirteen likes.

Mountain House, Age Eleven

GABE WHISPERS, "THIS IS THE PLACE." WE'RE IN THE north pasture, tall grass painting our jean legs with damp, morning sun coding the spider silk in bright dots and dashes. The spot where Gabe once spotted a western tanager, where we hope to glimpse another.

He slips a journal from his backpack and begins sketching. While he squints across the meadow, pencil twittering on the page, I leaf through my Golden Field Guide to *Birds of North America*. The same look and feel as my *Rocks and Minerals* book—a very '50s-era, pop-science aesthetic. A waxy, laminated cover, ruler printed on the inside back. Faint smell of glue on the interior pages. Illustrated birds crowded together against pastel backgrounds, their habitat range charted on maps in the left margin. To aid with in-flight identification, blue and black silhouettes migrate in formation across the top margins, wings swept wide.

*

Gabe kneels down, balances his sketchbook on a dry log, cranes his neck. Then bolts upright, binoculars at his face. Without looking away, he reaches for me with his right hand, pulls me over by the shirt, points to a stand of trees.

I do as Gabe taught me: follow his finger, scanning for motion and color and contrast until I find the right tree, then drive my binoculars up the trunk's vertical highway, before exiting the smaller branches, the leafier side streets where most birds settle. I crimp my neck, tuning the magnification with my right index finger, until I bring it into focus—my first western tanager—its breast brilliant yellow against dark pine boughs.

The face and head printed with vermillion. Golden epaulets on black wings above crosshatched white wing bars.

Like something you'd see in a pet store, or the tropics—flitting above Mr. Roarke and Tattoo in an episode of *Fantasy Island*—anywhere but a horse pasture in western Colorado. As the Golden Guide indicates, it is indeed a *tropical* bird that migrates north to breed in spring and summer.

With the tanager in sight, my binoculars are reverse headlamps, shining inward through my eyes and arteries and bones, down into the cavern of my small body, a brief burst of equatorial light touching bottom in my overgrown well of secrets. At the same time, these field glasses are prismatic conveyances, lifting me free from gravity's suck, momentarily transporting me skyward.

*

From the Golden Guide entry on tanagers: "Males often sing from the topmost branch."

*

On the walk back to the Mountain House, Gabe and I spot a northern flicker, a member of the woodpecker family. It clings to a tree trunk like a telephone headset in its cradle. Spotted underbelly, red mustache and a black bib, its call a shrill *chee-aw*.

Farther down the trail, tree-ribboned sunlight dries our boots, illuminates swirls of gnats and thistledown and bees scouting for pollen. Air thick with rusty musk from the aspen marsh, the scrub oaks' dry tang. We continue on through the wild strawberry and clover. To the east, horsetail and watercress buoyant on the creek.

*

That night, Gabe's mother, Sheila, reads *Huck Finn* out loud to us by the light of a propane lantern. She sits on the couch, calf crossed over knee. We sit cross-legged on the floor next to the wood stove, rapt by the story of a boy our age living out-of-doors. Such endless freedom, so many terrors.

*

Gabe and I were born ten days apart. Our mothers met six months after our births. We are the true definition of lifelong best friends, and I can't remember a time he wasn't drawing. Obsessively rendering landscapes through his right hand. Up in the cabin loft, using watercolors or making collages or charcoaling my small human figure on an oversize sheet of soft paper. Out on the deck or in the forest, plein air painting or sketching in hand-bound journals. His pencil and the outside edge of his palm schussing across the page, a sound like wooden skis on sun-glazed snow.

*

The Mountain House is filled with Gabe's art, and with books. Books everywhere—in the built-in shelves on the staircase, on the piano bench, in bunk beds and cubbies, stacked in the loft. One Friday afternoon I discover an old paperback copy of John Irving's *The Hotel New Hampshire* at the bottom of the stairs. Though it's wildly inappropriate for an eleven-year-old, I read it over the weekend, curled up in the loft while Gabe works on his first political cartoons. Addled by loss and misfortune, the characters in *The Hotel New Hampshire* always remind one another to "keep passing the open windows." It's here that I first notice the desire to become a writer.

*

Gabe and I sleep in lofted bunk beds, handcrafted by his father and always dressed with flannel sheets and piles of blankets. I take the top bunk, just below a skylight, reading from *Cosmos* by Carl Sagan, looking intermittently out at the stars—even using birding binoculars to magnify my view of the moon's pitted, glowing face. Looking out at the night sky—contemplating Sagan's concept of *infinity*—a sense of awe overwhelms the rock tumbler of my young mind. I wonder: Does the universe have an *edge*? If we had the right pair of binoculars or a proper set of wings, when would we glimpse that edge, and what would it look like? Just how far is it possible to expand outward?

Part Two

1.

IN THE DISTANCE, DOWN THE VALLEY, THE SKY BEgins to rip.

At first we think it's staccato thunder. A nightstorm moving up the mountains. Like the one that battered us a week earlier, salted our tents with an inch of hail. That previous evening, lightning strobed our campsite every few seconds, the sky cracking open and slapping shut, kick-drumming the ground beneath our sleeping bags. Our eight teenage students, whom my coleader Carrie and I are tasked with keeping safe during a month of backpacking in the remote Colorado mountains, huddled together for warmth and comfort under a blue tarp strung between two trees.

Not tonight, though. Above our campfire—above the coil of our adrenaline-shot young bodies—the sky appears shattered with stars. Not a single cloud between us and the fractured cosmos. No wind and little chance of a storm, just the smell of pine and spruce smoldering in our modest campfire.

The sound grows louder, the muffled rip morphing into a clipped *whip whip whip*. At some point it becomes recognizable: a helicopter. Of course a helicopter—it just didn't sound anything like one at first. As the rotor chop echoes around the forest, reverberates up the valley, it dawns on me that Carrie must have made the fifteen-mile hike down to civilization, called our employer at Silverton Expeditions, who in turn called Search and Rescue, and now this helicopter's here to help us find Alejandro—a sixteen-year-old student who has been missing for over twelve hours.

I should be relieved for the arrival of help, but all I can feel is dread.

We're on the tail end of our twenty-six-day trip in a program called *Wayfinding*. Modeled after Outward Bound, Wayfinding serves the oldest and most experienced students at the family-run Silverton Expeditions Camp based out of Durango. It's also my first job as a backpacking instructor after completing my own leadership mountaineering course in this same area with the Colorado Outward Bound School.

And, similar to an Outward Bound expedition, our Wayfinding students have just finished their three-night solos. Fanned out in a tight radius around Carrie's and my tent—never more than forty or fifty yards away from us—they spent three days resting and journaling. We checked on them multiple times a day. On the final day, when I went to fetch Alejandro, his belongings were all present.

Sleeping bag, backpack, water bottle, even his rain jacket.

But no Alejandro.

I assumed maybe he was just out for a bathroom break. We waited an hour. Two hours passed, then three and four.

We spread out in circles and shouted ourselves hoarse, his name echoing around the plateau, slapping off sheer granite faces to the west, then swallowed by the sky.

By early afternoon, a numbing sense of horror crept in. Especially after I rifled through Alejandro's backpack and located his journal, where his last entries described how he'd been wandering off by himself, rock climbing.

Reading this, my heart lost its purchase on the ladder of my ribs. Had I failed to fully explain the rules of his solo? That leaving his site was strictly forbidden? As was free solo rock climbing—the most dangerous conceivable thing he could do, our group a fifteen-mile hike from help. Alejandro was a Mexican citizen and though he spoke fluent English, I fear I may have failed to communicate across the language barrier.

Maybe it was the antidepressant medication at work, dulling my senses, thickening my thoughts and words with sludgy viscosity.

I scuttle to a high point near our camp for a better glimpse of the helicopter. From my vantage at nearly eleven thousand feet, I catch the faint glow of a spotlight, just beginning to illuminate treetops in the distance. All I can imagine: the beam slicing through the forest, landing on Alejandro's prone body at the base of a cliff.

A body only five years younger than mine. A body for which I'm entirely responsible, and that I'm terrified may no longer be alive.

Back at base camp, the campfire begins to register subtle shifts in air pressure. The flames warp and spark. I dump a half-empty Nalgene into the firepit. It hisses and sputters, but before I can fully extinguish the coals, the helicopter materializes right above us.

Embers and ash spiral into the dry forest.

The helicopter's sharp beam angles down on us, trained on these children who have hardly seen another person outside of our group, much less any kind of vehicle, for almost four weeks. We have to cover our ears, shield our eyes against this new weather system, this man-made storm.

As it descends to a flat spot, the blade whip torques and buckles the treetops. My focus narrows on one wind-burdened pine sapling with its tip curled almost to the ground, nearly snapping under the pressure.

The machine touches down a hundred feet from our campsite. As the rotor begins to slow, I creep toward it. A young man emerges, maybe thirty years old, wearing a fleece jacket and a chunky earphone headset.

"Did you find Alejandro?" he shouts, pulling one headphone away from his ear.

I share the bad news, explain what I learned from Alejandro's diary, that I fear he must have wandered off by himself and had a rock climbing accident.

"If he's injured," I say, "we don't have much time."

But he explains there's nothing more they can do, not in this darkness. That their landing here in our site is for information-gathering purposes only.

"You're leaving?" I ask, trying to mask the panic. Maybe even more than his competence, confidence, his actual *adultness*, I envy the fact that he'll be heading home to civilization this very evening.

He tells me they'll be back in the morning with seventy-five or more Search and Rescue volunteers. Many of the volunteers will have trained dogs, he says. We can expect at least three more helicopters, and a Navy jet will fly recon overhead.

The search area is staggering in size, he explains. They need all the manpower they can get, including me and our Wayfinding students.

"You and your crew need to be ready by dawn," he says, slipping his headphones back on. "That's when it all begins."

2.

IN THE SPRING OF 1992, AT A NEW AGE BOOKSTORE IN Sedona, Arizona, my father bought me a copy of *Iron John: A Book About Men* by the poet Robert Bly. I was eighteen or nineteen and had no idea who Robert Bly was, or that *Iron John* would go on to spend sixty-two weeks on the *New York Times* bestseller list and launch an entire movement. I had some familiarity with feminism, but I'd definitely never heard of a "feminist backlash." As a college freshman interested in the humanities, *Iron John* spoke to my growing love of mythology and the tracing of archetypal patterns across cultural narratives. It's embarrassing to admit now, but so did the book's central precept that, while feminism was generally a positive thing, many men in the orbit of strong, feminist women had become too "soft" and "tamed" after consistently bartering away their masculine power for vulnerability and consensus-based decision-making.

After reading the first few chapters, I remember wondering, *How do I measure up on the hardness scale? Am I too easily scratched?* Reading *Iron John* tapped a desire to reach the upper levels on the scale—maybe even to become diamond-hard, scratchable only by another diamond.

But the damage and relentless sense of anxiety I carried from childhood dropped my score way down to the 2 to 3 level, at least in my mind.

Maybe, I thought, *all the fear will go away if I can become enough of a man.*

To help young sub-men like me on our paths toward a brawnier gender expression, Bly suggests we get in touch with our inner "wild man"—a powerful, hairy, self-sufficient, Woodstock-era superhero who ostensibly dwells within every male's psychic man cave. And that we embrace our "Zeus-energy"—in other words, the ability to wield elemental power and authority in all the communities we occupy.

While hearkening back to the ancient Greeks, the mythopoetic men's movement, as it came to be called, also appropriated Native American cultural practices like sweat lodges and drum circles to help men cross the threshold from suppleness to dolomitic manliness.

On that same Arizona road trip, I found a hard copy of Joseph Campbell's *The Hero with a Thousand Faces* at a small library in the former mining town named Jerome. Knowing the library wouldn't lend the book to an out-of-state resident, I shoved it down my pants and walked out the door.

This seems about par for the course in my nascent sense of membership in the early men's movement. *Real men* take what they want—and stuff it in their jeans!—regardless of the consequences, in this case to what seemed like a cash-strapped public institution.

A couple months later, while reading Campbell's book on a camping trip with my mother and stepfather in Washington state, my mom approached me for a casual conversation. I was distancing myself from mother figures that year, as advised in *Iron John*. Instead of keeping her company, as she hoped for—and this memory makes me cringe—I explained that I was busy working on becoming a *hero*.

I remember her expression as she walked away. Bemused, disappointed, lonely, maybe even a little frightened?

On the way home from our Arizona trip, Gabe and I drove over Red Mountain Pass in the San Juan Mountains of southwestern Colorado. The most remote range in Colorado, they're hours away from any major city or airport and contain more alpine terrain above ten thousand feet than any other wilderness area in the state. Still quilted with patchwork snowfields, they transfixed me with their lofty and seemingly untrammeled qualities.

They appeared sufficiently rugged and isolated for what I was after.

Coasting along the top of the world with *Iron John* in my lap and my stolen copy of *The Hero with a Thousand Faces* on the dash, a vision overtook me.

In my fantasy a group of bearded men emerged from the mountains and fetched me away from my mundane life, taking me up above tree line—above all my worldly concerns and insecurities—to initiate me into true wildness and manhood.

What I didn't give much consideration at the time were the myriad tailings ponds scarring the landscape around defunct silver and gold mines up on Red Mountain Pass.

3.

THE PROCESS OF SEPARATING THE VALUABLE FRACtion from the worthless elements of an ore produces waste materials known as *tailings*. They're also often called mine dumps, culm dumps, slimes, tails, refuse, leach residue, or slickens. Generally contained in open-air reservoirs or pools near old mines, many are ominous shades of oxblood or jaundiced yellow. The poison lake of tailings above Butte, Montana, known as the Berkeley Pit, notoriously kills flocks of geese that touch down in waters with PH levels matching the gastric acids in a human stomach. Up the hill from the Berkeley Pit a 750-foot dam blocks 6.5 billion gallons of noxious soup from spilling out of the Yankee Doodle Pond onto the 34,000 Butte residents below.

Tailing containment failures are not uncommon. In 2015, beginning at a former gold mine near the site of my Red Mountain Pass vision, three million gallons of toxic slurry spilled into the Animas River, turning it sickly orange for days on end.

4.

RESEARCH CONSISTENTLY INDICATES THAT CHILDhood sexual abuse can have lasting effects on brain function, often leading to anxiety disorders, depression, disturbed adult relationships, and other complications. The amygdala—a deep brain organ involved in systems that regulate sex, eating, and internal threat response—seems disproportionately affected by abuse.

We might say that overstimulation of one neural pathway leads to outsize responses from other, connected pathways.

In my case, the mining of sexual pleasure from my young body led to decades of elevated threat responses, even in the absence of actual danger.

Low-grade anxiety plagued me as a child, but the real parasympathetic mayhem didn't come for me until I was sixteen. After using a fake ID to order five or six beers in a Tijuana nightclub, I drove back to East County, San Diego, and, though I opted out of dropping acid with my friends, smoked some dubiously sourced marijuana that, looking back now, I realize was likely laced with methamphetamines. Fortunately I was with my close friend Damian, who grew up twenty miles east, up in the Viejas Valley, and who was becoming more and more of a brother after moving in with my father and me on the heels of my dad's third divorce.

When I stepped out of the back seat of my car where I'd hotboxed the weed, my legs dissolved into the asphalt. "Quicksand!" I shouted to Damian and the others. I laughed it off and shambled after them toward the house of an upperclassman whose parents were out of town.

The minute I set foot in the small, suburban ranch house, I recognized a man's reedy voice emanating from the television.

"Is that fucking *Bob Crane*?" I said, audibly talking to myself.

My friends all converged around the dining room table, smoking weed from a crunched-up Coke can, discussing the ass-kicking highlights from a movie we'd watched the previous weekend: *Road House* with Patrick Swayze—a film that consecrated fist-fighting as the essential art form for any self-respecting man.

Over in the darkened living room, though no one was watching, the TV blared.

I shadowed my morbid curiosity and entered the empty room.

Where I stood on unsteady legs in front of the screen, squinting and blinking to make sure this was real life.

There, playing on the TV, was a laugh-tracked rerun of a poorly conceived sitcom I hadn't watched for a decade or more.

Hogan's Heroes.

On-screen, the Germans hold Hogan and his heroes against a wall at gunpoint, in preparation for their imminent execution.

The precise moment I feel my heart firing inside my chest.

And firing harder—and harder still—until it ratchets up to the rapid-fire pace of an automatic machine gun, matching the pace of Berettas *rat-atat-tatting* from the television—while I pace around the living room and kitchen, two fingers on my jugular, eyes on the ticking second hand of my watch, trying to count the by-now uncontrollable bloodthrum in my arteries and veins.

Can a heart really beat over ninety times in thirty seconds? I wonder.

Can a heart beat itself to mush?

I snatch a gallon jug of milk from the refrigerator and chug straight from the bottle, the upperclassman who lives in this house—and whom I've never actually met before—looking at me like *dude what the fuck?*

Almost perfectly synchronized with the peak of their acid trips, the shelling inside my chest sparks the realization that I'm having *an honest-to-God heart attack.*

The idea terrifies me, but my friends think it's hilarious, the way I've gone deathly pale and keep asking Damian to *please check my pulse.*

It's not my TV, but I switch the channel from *Hogan's Heroes* to a local news broadcast.

A deeply tanned male newscaster reports on the controversial spraying of Malathion pesticides—a neurotoxin known to harm the developing brains of children and teens—on crops east of our city, but all I can imagine is his polished voice recounting, the next morning, "the strange and awful case of an otherwise healthy sixteen-year-old San Diego teen who died last night of a massive coronary."

An agonizing hour passes, though it feels like ten.

I can't compel anyone to help me.

Lost in their own acid trips, my friends transform into demented, snickering toddlers.

The only one taking me halfway seriously, Damian gives me a harrowing ride to the hospital and escorts me into the waiting room.

An impatient ER resident walks me to a poorly lit exam room. He instructs me to lie down on a gurney and remove my T-shirt, so he can attach electrodes to my chest and monitor my heart rate.

Which now fires up near the two-hundred-beats-per-minute range.

"You're sure you didn't do any meth?" he asks for the second time.

"Just beer and weed," I reiterate, my eyes fixed on a metal wall clock that appears to be melting.

He shakes his head, tells me he has more important situations to deal with.

Alone in the room, with the intense rattling of my Gatling gun heart now digitally amplified, I reach a crescendo in the perverse, two-hour opera of my profound dread.

I writhe around on the gurney until I can't take it anymore.

Then sit straight up.

This is when the heart monitor—which until now sounded like the sped-up, final hundred-seconds countdown on a ticking time bomb—falls completely silent.

Flatlines.

Verifying that I am in fact dying.

"My heart stopped," I whisper into the lonesome, bleach-scented darkness.

I say it louder, but no one seems to listen. Or care.

"MY HEART STOPPED," I holler.

And then holler again, and again, until the resident returns.

He sighs and reattaches several of the electrodes that had come unglued when I sat up.

"Just relax," he says. "The Valium will kick in soon."

Mercifully, it does.

Though the sense of underlying dread lingered in my body and mind for years afterward. As did my obsessive compulsion to monitor my own pulse for signs of an attack.

The only benefit, other than making for an oddly humorous story to share decades later, was that it brought me to therapy, where the tainted secret I'd been holding all those years came spilling out.

It was an immense relief to finally confide in someone, to process what I came to understand as a series of assaults perpetrated on me. One of hardest parts was admitting that some of it had been pleasurable. That I was a lonely little kid and Clint gave me a warped kind of "affection." This led to questions about my own sexuality, as I tried to disentangle from the actual abuse my feelings of attraction to both women and men. Were my occasional crushes on men just trauma-related? Or an essential part of who I was?

Getting some professional help didn't mean the panic attacks ended, unfortunately.

I never knew when they'd hit—during a math test, the middle of the night. On an overnight flight or in a bathroom stall. While riding in cars or on chairlifts. Or in various ranch-style houses. They left me feeling hollow and stripped, like a chemical spill in my own bloodstream.

They felt so venomous and disabling that I developed a phobia of the attacks themselves. An intractable fear of fear—the exact thing JFK warned my parents and their generation against.

I grew increasingly hypervigilant in my anticipation of panic, attempting to control every aspect of my life to somehow circumvent the spontaneous over-massaging of my heart muscle with those frantic, icy hands.

A couple years later, in the garden-level University of Colorado counseling center, I tried more talk therapy, cognitive behavioral therapy, biofeedback, mindfulness meditation, immersion therapy for phobias. I listened to cassette-tape recordings of dulcet-toned practitioners encouraging me to breathe deeply while tensing and relaxing all the muscle groups in my body.

Finally I tried imipramine, an old-school antidepressant prescribed by a psychiatrist who spent all of ten minutes with me before writing out the scrip.

The therapy helped some, while the imipramine mostly just made me feel doped up and dull-edged.

Neither provided the immunity from panic I so desperately craved.

5.

AFTER THE SEARCH AND RESCUE HELICOPTER FLEW off into the night, I had nothing left, emotionally, to keep morale up among the Wayfinding students.

Instead I sat alone with my spiral-bound journal. Under the light of my Petzl headlamp, I wrote my name over and over, Jack Torrance from *The Shining* style, to avoid flying off the face of the planet.

My name is Justin Hocking. My best friend is Gabe Liston. My name is Justin Hocking. My best friend is Gabe Liston. My name is Justin Hocking. My best friend is Gabe Liston. My name is Justin Hocking. My best friend is Gabe Liston. My name is Justin Hocking. My best friend is Gabe Liston. My name is Justin Hocking. My best

Then I lay petrified in my sleeping bag, waiting for dawn and the arrival of more helicopters.

And for the crippling panic to sink its teeth deeper into me. Exacerbated, now, by the thought of how frightened Alejandro must've been, alone out there in the pitch black.

If he was even still alive.

My ribs ached for him and his family.

And it felt clear that my secret was about to be revealed.

I'm a boy on medication for an anxiety disorder masquerading as a capable backpacking guide and mountaineer; I'm a damaged child masquerading as a man.

In the excruciating court case already well underway inside my head, jurists had been selected by my poorly dressed public defender and the polished, high-powered lawyers hired by Alejandro's wealthy parents.

A court case wherein my secret would not just be revealed, but also used against me, as a causal factor in my clear liability for his disappearance and death.

Which meant the disastrous end not just to my budding career as a wilderness guide—and whatever mental stability I did have—but to the entirety of Silverton Expeditions, a family-owned outfit whose management passed down through generations since the 1960s.

6.

NOT LONG AFTER MY FATHER PURCHASED THE COPY of *Iron John*, I picked up a Colorado Outward Bound School catalog from an informational booth on my college campus.

Flipping through the catalog, I was elated to find they offered thirty-two-day leadership mountaineering courses in the San Juan range.

"It is home to many abandoned mining towns whose rather checkered pasts lend credence to Colorado's 'Wild West' reputation," the catalog copy read, further piquing my interest.

At that point the question of my sexual orientation was still unresolved. Enlightened as she is, my mom made it clear that she'd accept me either way, but that she hoped I wasn't gay, because "it's such a difficult life." My dad remained pretty much silent on the subject. I knew I liked women, but in my freshman year of college I also developed a crush on a boy in my French class. He was so sweet and uninhibited in his femme-ness, like no guy I'd ever met before, always reciting our French lessons with his intentionally outrageous accent—smiling at me and saying, "*Voulez vous couchez?*"—and then laughing at his own jokes.

I didn't act on it—I had no idea, at the time, how I'd explain it to my friends.

Instead, I kept reading *Iron John* and fantasizing about Outward Bound.

While many of my college friends spent the summer after our freshman year playing in punk bands or skateboarding—or in Gabe's case, honing his visual art skills—I spent mine working on a construction crew, schlepping around stacks of three-quarter-inch plywood, cutting and nailing together headers for doorways and windows. During brief lunch breaks I went home and watched reruns of *Bonanza* or a VHS copy of interviews with Joseph Campbell and tried to puff myself up for the remainder of the workday. Even so, a couple of the jaded framing-crew lifers always gave me a hard time. I was strong and had good hustle, they said, but I didn't have the right instincts to work construction.

In other words, I was only *book smart* and *not man enough*.

If I had to do it over, I'd probably work in a restaurant or intern at the local newspaper. Though I did pick up some invaluable carpentry skills, and earn enough money to pay my own tuition for the Outward Bound course the following summer.

Early one June morning in 1993, Gabe drove me along the mist-feathered Colorado River from Glenwood Springs to

Grand Junction. There I boarded a yellow school bus for a ride south, over Red Mountain Pass—the original site of my *Iron John*–inspired vision.

The other fledgling Outward Bounders and I marveled out the bus windows at the San Juan range's grandeur. We also laughed nervously about the way that guardrails apparently weren't in the budget for the winding, vertiginous pass, even though it was nicknamed the "Million Dollar Highway."

The bus dropped us off at the top of Molas Pass, at 10,900 feet.

This is where we met our instructors—mostly bearded, hardy, incredibly fit men in their late twenties or early thirties.

Just as I imagined.

During orientation, we learned some basic rules and skills. For the next month, we'd carry all our own food, water, and supplies in old-school canvas backpacks that, when fully loaded, weighed as much as sixty or seventy pounds.

They showed us the orange metal trowel we'd use to dig our own latrines. For whatever reason we nicknamed it "Stan." Along with topographic maps and compasses, they gave us each a little amber bottle of iodine tablets to treat creek water.

Though they advised against drinking water from the Animas River, due to mine tailings.

I'd grown up in Colorado, but at the time I didn't know exactly what mine tailings even were. And later in the course, I heard whispers from our instructors that, because of all the unearthed heavy metals, elevated incidences of testicular cancer existed for men who lived in the area.

It'd been a heavy winter for snow, so we carried ice axes and, within the first few days, learned self-arrest techniques on snowfields, which meant hatcheting frantically on the ice after a fall, praying for enough purchase to keep your body from sliding off a cliff face or into a crevasse.

I was surprised to hear one student on the course had been sent by his parents. He was working on getting his life together after a couple stints in rehab, and wasn't particularly enthusiastic about getting shipped away again. Nor was I enthusiastic about his carping attitude, having planned and worked an entire year for this experience.

The course was so demanding that he lasted all of five days.

On the other hand there was Mark, a couple years younger than me, who had already enlisted in the Marines and was leaving for basic training shortly after our trip. One of our instructors, also named Mark, was a former Marine himself, though you might not guess it based on his lithe frame and serene nature.

As instructor Mark explained to student Mark, aspects of our thirty-two-day leadership mountaineering course required more physical endurance than boot camp for Marines.

However hard it was, I fell into a natural leadership role as our first expedition unfolded. On the trail, Mark and the others liked having me up front, where I kept a decent pace and felt fairly adept with navigation via map and compass. I was also one of a handful of students actually from Colorado and pre-acclimated to the thin air.

The younger Mark and I became fast friends; I was flattered that he seemed to look up to me.

Another fellow student worked as a chef in his hometown of New Orleans. After one particularly grueling day backpacking over a 12,500-foot pass between inscrutable cathedrals of vertical granite, he stretched out his foam sleeping pad and used his Nalgene water bottle to roll out tortillas from scratch. He wrapped them up into tidy bean-and-potato enchiladas, dripping with melted cheddar and garnished with wild onion he sourced from a nearby meadow.

The next morning I commented that his Cream of Wheat was "the perfect sweetness." Everyone laughed, and the moniker ended up sticking as our crew's official name.

More than a nickname, it embodied everything about my experience for thirty-two straight days. I adored the other students, our instructors, the San Juans as they molted their snowfields to reveal verdant, rough-hewn summer skin.

For me it was a month of exertion-fueled bliss.

One morning, while camped in the snow around Trinity Lake, our instructors woke us up at 3:30 a.m. for a summit attempt on Storm King, the peak whose sheer north face loomed like a haunted castle three thousand vertical feet above the icy blue tarn.

The stars were fading pinpricks in the indigo dome of sky. Our headlamps cast their own frail points of light as we peered up at our formidable route along Storm King's east ridge. A cold, still morning—the only sounds the crunch of our boots on glazed, hardpack snow and the occasional rumble of loose boulders beneath our feet.

With help from lines fixed by our instructors we summited Storm King by noon.

And then stood triumphantly on top, brandishing ice axes like swords for a group photo.

Our enthusiasm was short-lived, however.

We scrambled down Storm King's south face and circled around to our base camp for a brief lunch, only to be informed that we had to pack everything up and slog six miles to our next campsite at Balsam Lake.

Balsam Lake, on the south side of the same 12,500-foot pass we'd just descended.

We were already exhausted and would now have sixty-pound packs weighing us down.

As we stood around grumbling, I made a joke at the expense of the more serious and driven of our two instructors—the one who led the charge in the unpopular decision to relocate our camp. He was a good-looking guy, but with his round spectacles and Germanic features he faintly resembled the ruthless Gestapo agent in *Raiders of the Lost Ark*, the one who sadistically interrogates the female lead, Marion. And whose face melts off dramatically after he and his Nazis compatriots open the Ark of the Covenant.

When I pointed out the resemblance for my fellow students, Mark followed up with the slam dunk.

"Ve haf veys of making you *hike*," he whispered in his campiest German accent.

The full-belly laughter cheered us up. At least until we remembered the task ahead of us.

By consensus we chose an amiable, well-read student from North Carolina to lead us over the pass. He was a soulful person who liked road-tripping on his vintage motorcycle, and who could talk for hours in his southern drawl about his love for Mark Twain.

With so much weight on his back, his size-twelve boots kept post-holing two or more feet down in the slushy, late-afternoon snow. Less than an hour in, and with both legs sunk up to his thighs in snow, he started bellowing like a stuck moose.

Mark and I had been quietly taking up the rear but were now requisitioned to move up into the lead spots. With Mark shadowing me I methodically kicked steps in the snowpack, building a narrow, switchbacking staircase up the precipitous slope. Whenever I post-holed, Mark grabbed me by the backpack and hoisted me up with his brawny, pre–Marine Corps arms. It was slow going but everyone seemed to appreciate our efforts.

By 7:00 p.m. we summited the pass in a snowstorm. Rejoicing even in the June blizzard, we couldn't help but feel heroic.

Mark and I took up the rear again on the more gradual descent. Deeply fatigued, we fell behind and didn't arrive at our new camp until well after dark, after a full sixteen hours of climbing and hiking.

Thankfully our other group members still had the energy to set up our tarps and cook the most sublime macaroni-cheese-and-tuna casserole I ever tasted.

Weeks later, I wrote the following in my post-course evaluation:

> The high point of my course was reaching Balsam Lake and a beautiful green meadow after climbing

Storm King and then hiking all day—the hardest physical day of my life . . . I was exhausted by the time we made it to that incredible campsite. It was one of the best feelings of accomplishment I've ever had.

Most people sit around never exerting the energy we did that day. They might be more comfortable, but they will never see life's green meadows.

And then, a year later, I was elated to find this exact quote, attributed to me, in Outward Bound's promotional catalog. My very first piece of published writing, placed at the top of a page in a dark green, bold-faced font.

Though looking back now, I have serious misgivings about the ideology behind my words.

To be sure, the trip was quite possibly the most exhilarating month of my life.

But, along with the overreliance on passive-voice verb constructions—something I always nitpick writing students about—I feel uncomfortable with the ableist sense of superiority I read into my sentiments, and with my positioning of the backcountry as a kind of blast site for my *seize the day* instincts.

For my personal accomplishments.

Most problematic is the last line, wherein there's a clear condescension, bordering on arrogance, toward all those people lacking the fortitude to reach alpine meadows up at eleven thousand feet.

I can sense Robert Bly's hand in my prose, disapproving as he was toward "soft" men.

I detect similar sentiments in former Outward Bound Executive Director Brady Udall's introduction to the 1994–95 catalog. He begins by asking the reader if they've ever had a real adventure "where you could stand up in front of family, friends, teachers, bosses, or coworkers, and know in your heart that you had accomplished something extraordinary? . . . And maybe, just maybe, astound yourself and others with bravery, perseverance, compassion, confidence and strength."

I don't want to disparage the remarkable people who lead Outward Bound trips, and all the hundreds of thousands of young folks the program has touched. But both quotes epitomize Outward Bound's emphasis on character building over care for and communion with nature, at least at that time.

We can trace this ideology all the way back to Kurt Hahn, one of Outward Bound's original founders. Hahn courageously stood up to Hitler before escaping from Nazi Germany. In England, he founded the Gordonstoun school based on the noble aspiration that holistic education might create a better society and deter another rise of fascism. Some have described Hahn's philosophy as "education for democracy."

Nevertheless, others criticized Hahn for his insufferable arrogance, and for wielding an iron fist against his teaching staff—none of whom, subsequently, gave him their full adulation or respect. During his lifetime he formulated his pedagogical beliefs into what came to be known as the "Six Declines of Modern Youth," including his spurious claims that there existed, among young boys especially, drastic "Declines in Fitness" and "Declines in Initiative and Enterprise," due to the "widespread disease of spectatoritis." As well as alarming "Declines of Self-Discipline," due to the "ever-present availability of stimulants and tranquilizers," by which I assume Hahn meant coffee and beer.

Or, if they'd been around in Hahn's era, antidepressants.

To counter these so-called declines, Hahn prescribed, among other things, rigorous programs of Fitness Training and Expeditions—the core elements of which I experienced that day on and around Storm King.

Given the affinities between their concepts of "decline" and "softness," it's easy to imagine Hahn and Bly sharing their contempt for effeminate, nonathletic boys while vigorously chopping cordwood or grouse hunting.

To be fair, the '90s-era Outward Bound catalog does devote one page to their ethics of environmental stewardship, and their belief in the potential for young people to develop deep connections with nature. But overall the catalog's messaging tends more toward the benefits of impressing people with your superhuman triumphs.

And, with a rhetorical boost from my own quotation, placing yourself way up high, ten-thousand feet above all those average plebes and lowlanders.

Looking back now, I understand this as a precarious position.

Or, more precisely, a dangerous position for my own self.

When I reflect on the phrase "Outward Bound" as a middle-aged adult, dozens of positive associations arise. As a writer, I also can't help but consider other, more constrictive connotations of the word *bound*.

Toward the end of our month-long trip, we all completed our three-night solos. I fasted the entire time, which increased my sense of vulnerability to the elements.

On the first night, a thunderstorm cleaved open the night sky.

I lay beneath my meager tarp stretched between two fallen logs, reading *Leaves of Grass* under the light of my headlamp. When heart palpitations kicked up and cortisol flash flooded my veins, I seriously considered abandoning my solo camp to seek refuge and succor with my instructors, who were camped less than a quarter mile away.

But I breathed through the panic until it eventually passed. As did the thunderstorm.

I awoke the next morning before dawn, ecstatic for the pale light just starting to edge over the Trinity Range to the north.

Ecstatic, too, that I'd breathed and talked myself through an anxiety attack.

Meaning it no longer had its claws in me.

Meaning I'd conquered my fear.

Overcome my softness.

After solos, we reconvened for a collective sauna, constructed for us by our instructors with spare tarps, climbing rope, and rocks.

With this humid, literal initiation under my belt, I came home from Outward Bound feeling nearly indestructible.

I told my girlfriend at the time I was in Superman-level shape.

For maybe the first time in my life, I no longer felt like I was living belowground.

In my mind, I'd passed the hero's journey with flying colors. Now I was ready to show others the way.

And the only way to do that, I believed, was to become an Outward Bound instructor myself.

On the very last day of our trip, during a debrief with my instructors, they said that though I was the model OB student, I now knew enough to get myself in serious trouble. "The hardest part for you," they said, "will be now that it's over."

7.

BY DAWN THE DAY AFTER ALEJANDRO'S DISAPPEARance, helicopters swarmed the sky.

Four or five of them.

Including a camo green, double-rotored Chinook helicopter, operated by members of the Army Reserve. It looked massive and unwieldy in the air, like a robot bumblebee or a deep-sea submarine hovering on pinwheels.

The Chinook and other aircraft ferried scores of Search and Rescue volunteers up from what they called the command center down near Durango.

As promised, a Navy jet flew reconnaissance sorties over the mountains. Though it cruised at such a high altitude I wasn't sure of its utility, other than to slash the sky open every thirty or forty minutes, spooking me and our already-rattled Wayfinders.

The whole operation was, as I'd later learn, the largest Search and Rescue campaign in La Plata County's history.

The closest thing to a combat zone I've experienced.

The Search commander was a weathered, no-nonsense man in his forties. He took me aside and plied me for information, but I didn't have much to offer beyond the basic details.

I could still force my body to ambulate around the woods. But emotionally I was close to catatonic with dread. An automaton in hiking boots and a fleece vest.

I kept rambling on to the Search commander about how I assumed Alejandro was probably at the bottom of a cliff somewhere.

He regarded me with barely concealed pity. Or was it contempt? Or some combination of both, like I imagined Robert Bly or Kurt Hahn might judge me at this point?

The Seventh Decline of Youth: A pathetic, unmanly inability to keep track of one's students in wilderness settings, or, in the case of Search and Rescue operations for said students, to effectively communicate with one's superiors.

Eventually the commander assigned me and the Wayfinders to various search crews. A couple other Silverton Expeditions counselors and their younger students happened to be in the area, so they were engaged too. As was my close friend Nick, the senior Silverton Expeditions rock climbing instructor.

Mostly we wandered around in circles through the mosquito-infested highlands—located twelve or so miles, as the crow flies, from a ski resort known as *Purgatory*—almost getting lost ourselves and confronting the harrowing awareness of how easily this dense forest can scramble your sense of orientation.

At one point, Nick and I found ourselves in the same search cohort. Earlier that summer, during a two-day break between the first month-long Wayfinding session and our current one, he and I went out for drinks. Out on the back patio of a Durango cowboy bar, I opened up about what a hard time I was having, how lonely I felt out in the field, even though I was surrounded by students and had no time for myself.

Fighting back tears, I told him how, up there in the mountains at night, I'd been waking up with severe anxiety.

Nick was a much more experienced climber and outdoorsman, three or four years my senior. He was originally tapped as Carrie's coleader for Wayfinding, but he didn't relish the idea of spending his entire summer—nearly sixty straight days—with students in the field. Instead, though I was only twenty at the time, I was hired to colead Wayfinding, while Nick subcontracted as a kind of freelance rock climbing instructor. Twice during each Wayfinding session, he met us in the field to fix our top-rope gear and supervise our student climbers, before hiking back to his own life.

At the Durango cowboy bar, Nick really had no response to my disclosure about the anxiety disorder.

In his silence I read a declining faith in my ability to lead children in the mountains.

Now, wandering our quadrant of the Search and Rescue zone, Nick did his best to keep my spirits up. "He's probably just out playing with his nuts," he said, echoing the ridiculous tradition among this session's Wayfinder boys to insert every male group member's first name into their juvenile earworm of a song:

Alejandro, the Squirrel, playing with his nuts.
Justin, the Squirrel, playing with his nuts.
Nick, the Squirrel, playing with his nuts.

Beneath Nick's outward levity, did I detect the sense that he knew this would happen?

In comparison with men like Nick and the Search commander, was it possible to score *less than zero* on the Geologist's Field Hardness Scale?

I hardly slept that night, or really any night in the many days to come.

The next morning the Search commander took one look at me and the Wayfinder students and pronounced that it was time for us to evacuate the plateau.

I made a case for us hiking back out to civilization.

He insisted we go by helicopter.

Following orders, we packed up our things and stumbled to the makeshift heliport, located on a grassy meadow in a tight semicircle of trees.

While waiting anxiously for the Chinook, we watched a small, two-seater helicopter attempt to take off.

Whereas the other crafts seemed to lift off and swiftly clear the treetops, this pilot struggled to gain the necessary loft. Moving forward more than upward, he was forced to bank a sharp turn around the edge of the forested semicircle, his rotor blades narrowly missing the trembling pine boughs in his path.

The other volunteers and pilots stood around the meadow, grimly watching the scenario play out. A rescue-dog trainer's hand flew up to cover her shocked expression as the helicopter began losing altitude.

My hands on my head, I couldn't help but picture the helicopter's fiery demise—the rotor axing violently against thick pine trunks, transferring all its torque to the vulnerable cockpit, where the pilot would surely die—yet another death I could blame on myself—upon corkscrewing impact with the earth.

Though his craft was literally falling from the sky, the pilot somehow thrusted forward, finding a narrow path through the forest with enough forward momentum to finally punch out over the mountainside.

Around the meadow we let out a collective *whoop* of relief, but the sound of the much-louder Chinook drowned out our brief celebration.

Busy watching the struggling pilot, we didn't notice the Army helicopter in a holding pattern nearby, preparing to land.

Kicking up a torrent of dust and pine needles, the Chinook set down with all the grace of an airborne walrus. Then fully uniformed Army Guardsmen helped us collect our backpacks and board through the helicopter's gaping back end, where they cinched us to bench seats along the walls.

One of the Wayfinders, Aria, clenched my hand in hers the entire ride. Truthfully, I needed her grip just as much as she needed mine. The Chinook's back end remained wide open for the flight's duration, allowing us an unobstructed view down to the search zone. From this precipitous vantage, my entire sense of scale shifted for the landscape below.

The search zone was just an insignificant bubble in what I now perceived as an interminable ocean of forest below.

There's no way we'll ever find him, I thought.

Not in that endless evergreen sea.

8.

Common Substances Found in Mine Tailings:

- Arsenic
- Barite
- Calcite
- Mercury
- Radioactive materials
- Sulfur
- Hydrocarbons
- Cyanide
- Sulfuric acid

9.

Minimum Skill Requirements for Becoming an Assistant Colorado Outward Bound School Instructor:

- Five fourth- and fifth-class peak ascents spanning at least two seasons
- Familiarity with snow-climbing techniques (use of ice ax and snow anchors)
- Off-trail route-finding ability in mixed terrain
- At least thirty days of extended wilderness expeditions spanning at least two seasons (winter and summer)
- Ability to comfortably lead at least a 5.6 standard rock climb
- At least ten multi-pitch fifth-class climbs in past two years
- Familiarity with belay/rappel techniques, brake systems, and cliff evacuations

10.

AFTER MY IDYLLIC OUTWARD BOUND TRIP, I NEVER wanted the feeling to end.

I hoped to stay as far aboveground as humanly possible.

High alpine terrain—or *life's green meadows*—being the only place I thought I could feel so fully alive and happy.

Within the pages of the same Outward Bound catalog that published my quotation, I located the inventory of minimum requirements necessary for becoming an assistant instructor.

Though it intimidated me, especially the requirement of ten multi-pitch climbs, I fixated on it.

Everything I did, I did for the list.

With a group of new friends from the University of Colorado Outdoors Club, I summited a fourteen-thousand-foot peak in the dead of winter. And then, on the descent, with the harsh realization that I wasn't entirely immune to altitude sickness, I vomited in the snow.

The next summer I landed a job working with early teens at Silverton Expeditions, where I hit it off with Nick, and worked my way up the ranks. And was eventually offered the Wayfinding position for the following summer—a rare opportunity after only one season as a guide.

Wayfinding wasn't Outward Bound, but the differences were fairly negligible. Wayfinders spent the entire summer backpacking in the San Juans, with occasional expeditions in the exceptionally remote Dark Canyon area of Utah.

One contrast was that we didn't take our high-school-aged Wayfinding students on technical, multi-pitch ascents of fifth-class peaks.

So in many ways, I saw Wayfinding as just another stepping stone in my true path toward employment with Outward Bound.

And in my life as a mountaineer.

I pushed myself harder and harder to check items off the list during the year before my Wayfinding stint began. Everything else fell to the wayside: bird-watching, relationships, schoolwork, self-care.

One September afternoon in Boulder, I joined my roommate Remy for a push up the Third Flatiron—one of the most classic rock climbing routes in the world. It's essentially a giant slab that poses few problems for most climbers. Many do it without ropes; one joker even summited in full scuba gear.

Still, the east face of the Third Flatiron looms nearly 1,200 feet high, and climbers usually scale it in six to eight pitches—a *pitch* roughly representing one full length of a climbing rope.

My roommate Remy led the ascent. It was easy climbing on beautiful rock—an orange-hued amalgamation of sandstone and feldspar. Though I felt edgy from the very beginning.

At the end of each rope length, Remy set up a new belay station for me. Meaning that, while he climbed, I hung back with my weight on my climbing harness, giving me the proper leverage to belay. After the third or fourth pitch, and fully strapped to the cliff face, I made the mistake of looking down.

At six or seven hundred feet up, I realized that I was, more or less, perched on the ledge of a slanted skyscraper.

My father was afraid of heights, and I'd always been afraid, to some degree. I'd kept it together on Outward Bound—and actually enjoyed certain exposed climbs, like Storm King, because I was in the presence of exceptionally well-trained adults.

But up on the Third Flatiron, the panic stung my heart with the same force as it had in that suburban ranch house, back in San Diego. The difference being that now, four or five pitches in, I was fully committed to this climb.

There was no easy way off the ledge.

With trembling hands and a chest full of hornets, I worried about my ability to properly belay, or to keep my grip on the rock. Seven hundred feet in the air is never the best time to realize there may be a fundamental incompatibility between your psychological constitution and your choice of leisure-time pursuits. Especially when you've built your entire sense of purpose around said pursuit.

I really, really did not want to continue scaling this building-made-of-stone as much as I felt obligated—I needed to keep *standing up and showing people what I'd accomplished.*

To keep myself from tumbling back into the depths.

For whatever reason, the song "Africa" by the '80s band Toto lodged in my mind. I sang it out loud in my terrible, off-key way—doing anything I could to distract the frantic animal inside me.

After another agonizing half hour, Remy and I summitted just in time for sunset.

While he took in the sight of the surrounding hills and the city of Boulder stretching toward the eastern horizon, I could not get over the fact that we were standing on the narrow roof of what was only a few hundred feet shy of the Empire State Building—that is, if the Empire State listed heavily to one side, like the Tower of Pisa.

If any of the resident red-tailed hawks or violet-green swallows soared near us, I was too scared to notice.

Relief washed over me after we rappelled off the Flatiron's backside and then hiked down to our car in the twilight.

Though my nervous system felt shredded.

Especially after Remy informed me that the Third Flatiron hardly qualifies as a bona fide fifth-class climb. Which meant I still had ten steeper, more difficult multi-pitch climbs to cross off my list. And not as a follower, but as the lead climber.

A couple weeks later I attempted a shorter but more technically challenging two-pitch climb with Remy up in Eldorado Canyon. And then, the following day, during a math test in a poorly lit lecture hall, another anxiety attack pummeled me, seemingly from out of nowhere.

Within just a month of the Flatiron ascent, my fear of fear returned with a vengeance.

The panic and phobias spilled over into nearly every aspect of my daily life.

It got to the point where I couldn't ride a ski lift—something I'd been doing since I was a toddler—or get on an airplane without the company of a "safe" person like my girlfriend or my mother.

If only Robert Bly could see me now, I thought.

To make things worse, a subsequent crash in my mood almost always followed the intense spikes of cortisol and adrenaline in my system.

My mother labeled them as dysthymic bouts of depression.

With my Wayfinding instructorship set to start in just a couple months, my girlfriend and I broke up in the spring, sending me spiraling into an even deeper melancholy.

On the phone with my future co-instructor, Carrie, while sipping water to mask the dry mouth from the recently prescribed imipramine, I tried to sound excited for the summer.

11.

THINGS DIDN'T GET ANY EASIER ONCE THE CHINOOK dropped me and the Wayfinders back in civilization. We were disconcerted to find the command center helipad ringed with news vans, camera people, reporters. As we exited the helicopter they shoved microphones in our faces, demanded answers to their questions.

"What do you think happened to Alejandro?"

"Why was he left alone in the wilderness?"

"Do you think they'll find him alive?"

Back at Silverton Expeditions base camp, I learned that Alejandro's parents had flown in from Mexico. Though she wasn't in the search zone, Carrie's experience of trying to explain the situation to them had been no less difficult or traumatic.

During a session with a trauma counselor from Durango who seemed a little out of her league, I broke down in front of her and all the Wayfinders, repeatedly saying, "This is all my fault."

Later that night, side by side in our sleeping bags on the dusty shag carpeting of the camp's administrative office, Carrie and I promised each other that if Alejandro wasn't found—or if he was dead—we'd switch careers to accounting—or any other field in which we'd never again have responsibility for someone else's life.

We waited for two more days, then three.

Still no news.

It was late August, with fall and then winter approaching.

Up at ten thousand feet, the nights grew bitterly cold.

Colder than anyone could likely survive without a sleeping bag.

Without even a jacket.

We received word from the command center that, after five or more days of fruitless searching, they were close to calling it all off.

The mood around camp grew severely grim.

We had nothing to do but wait.

Wait and watch the Wayfinders' expressions fill with increasing despair.

The same despair etching its way onto the camp owners' faces: *this might be it.*

The last of thirty or more summers for their family-run business.

Alejandro's last summer on Earth.

12.

MOST MEMBERS OF THE 1990S-ERA MYTHOPOETIC men's movement claimed to be apolitical. Rather than siding with the left or the right, they ostensibly focused on men's psychology and wellness.

But their ideas left some traces on the cultural and political landscape.

Or tailings, if you will.

To be fair, Robert Bly and the first-wave men's movement were deeply critical of toxic masculinity, long before the term entered our everyday cultural lexicon. Bly believed too many men hid their insecurities behind puffed-up masculine disguises. And perhaps, in anticipating what further dangers might arise without organized interventions against toxic masculinity, he sagely and quite accurately predicted the perils of the Trump era.

Bly and other early men's movement figures failed to anticipate, though, that metastasized versions of their own gender-essentialist ideology—the notion that "men need to reclaim their essential status as men," as gender theorist Julia Wood described it—would actually help fuel the unchecked cellular growth of the more virulent second and third wave men's movements, now splintered into groups like involuntary celibates (incels) and militant, neofascist hate-mongers the Proud Boys.

In the lead-up to the 2016 presidential elections, men like Canadian YouTuber Jordan Peterson began decrying a crisis of masculinity and claiming that the masculine spirit is under assault. By recycling Bly's ideas—with help from YouTube's greed-driven algorithms designed to command the most possible viewing time, despite the questionable ethics of much of its content—Peterson and his ilk have contributed to the radicalization of thousands of men online.

The podcast *Rabbit Hole* chronicles one such young man who, catalyzed by Jordan Peterson's grievances about the supposed victimization of white men, transformed from a left-leaning punk rocker to a full-on alt-right Trump supporter.

In author Raewyn W. Connell's book *Masculinities*, she theorizes the existence of a quagmire with regard to Bly's construction of masculinity in *Iron John*—a construction that is, in turn, largely based on psychologist Carl Jung's conception of the Anima and the Animus archetypes.

According to Jung, the Anima and Animus are the essential feminine and masculine elements of our psyche that require balance in order for an individual to attain full actualization. In other words, Jung believed men do need to get in touch with their feminine side, so to speak.

However, Connell makes the observation that, according to Jung and his literary descendants like Robert Bly:

> No historical change in their [men's] constitution is conceivable; all that can happen is a change in the balance between [the Anima and Animus] . . . in modern Jungian writing this yields an interpretation of feminism not as resistance to the oppression of women, but as the reassertion of the archetypal feminine. In past history it is not men who have dominated women, so much as the masculine that has dominated the feminine. One can see why Jungian theory has become central to the current backlash among formerly progressive men. For this approach immediately yields the idea that modern feminism is tilting the balance too far the other way and suppressing the masculine. Bly's influential criticism of "soft men" who have caved into feminism and thus have lost the "deep masculine" is based precisely on this Jungian formula of archetypal balance.

Connell goes on to write that, in terms of gender theory, "Jung turned his back on a socially literate psychoanalysis that was pointed out by Adler and Horney."

Perhaps the deepest flaw, then, in the original men's movement was the positioning of feminism as the competition—or in the eyes of the Proud Boys and most right-wing conservatives, *the enemy*—rather than as a source of elegant solutions to so many man-made catastrophes we now face.

At their core, were Bly's ideas an ethos of narcissism, as Michiko Kakutani claims in her blistering 1993 review of *Iron John* for the *New York Times*? At the very least, it's safe to say that *Iron John* and its accompanying mythos have not aged well.

As for me, Bly's influence led to some unforgettable adventures in my youth, filled with wonderful friends and teachers and stunning landscapes.

But ultimately my rejection of my own "softness" was a way a way of punishing myself and stripping the joy from my life.

13.

WHEN PEOPLE BECOME SERIOUSLY DISORIENTED IN THE wilderness, they almost always wander downhill, though they may not be aware of doing so.

They lose elevation, in other words.

In the early morning of the devastating fifth or sixth day of the search for Alejandro, after hundreds of volunteer hours and well over two hundred thousand dollars in expenses—and just before calling the whole thing off—an idea occurred to the same young man who flew up that first night in the helicopter.

He guessed that perhaps Alejandro had wandered *way* downhill—farther than anyone imagined.

Even as far as the formidable cliff bands five or six miles south of our campsite.

Later that same morning, the Wayfinders, Carrie, and I sat around in the camp cafeteria trying to force down another depressing breakfast of reconstituted eggs and cereal from plastic bins.

Several Wayfinders considered going back home; they were already discussing flight arrangements with their parents.

Then, from the corner of my eye, I saw a camp administrator's pickup truck roar into the parking lot in a torrent of dust.

He burst into the cafeteria and from his jubilant expression alone, we knew.

As we all stood up in anticipation, the camp administrator fell to his knees.

"They found him," he shouted, "and he's ALIVE."

We all crumpled to the floor, shouting and hugging and crying for joy.

At that moment, Alejandro was already in a helicopter, on his way back to us.

They'd located him at the base of the cliff bands, huddled under a rock outcropping.

A scared boy just barely surviving inside a cave.

He was in good health but weak enough from malnourishment and exposure that they decided to drop a rescue litter from a helicopter and hoist him off the cliff bands, then deliver him directly to the hospital.

After hospital staff got him rehydrated and released him with a clean bill of health, his parents brought him back to camp.

This sixteen-year-old boy now looked like a forty-year-old man.

As if he was Carrie's and my senior.

He was pale and gaunt but in good spirits.

As he explained it, the last evening of his solo he wandered off to watch the sunset from a butte near his campsite. Then, walking back in the near-dark, he must've just barely missed his site.

He'd spent the rest of the night stumbling around, calling out for us in vain.

A dozen or more hours later, he located the outcropping, where he finally took shelter and decided to stay put.

He survived in that cave for five days, T-shirt pulled over his knees for warmth, living off fetid pools of rainwater. And eating the occasional mosquito or ant, just like we'd taught him.

A scared boy, huddled inside a cave.

Despite everything he'd been through, he made the firm decision to actually finish the last week of Wayfinding with us, rather than flying home to Mexico with his parents.

And years later, he became a Silverton Expeditions Camp counselor himself.

Sitting together on the wooden deck outside a camp cabin, Alejandro told me he was well aware that he wasn't supposed to wander off by himself, that he knew he was breaking the rules.

"Everything that happened here," he said, "none of it was your fault."

14.

IN THE CLASSIC UNFOLDING OF AN ARCHETYPAL HERO'S journey, after a descent and a long series of ordeals, the hero discovers his "treasure" and returns home to share the riches with his community. The wealth is generally spiritual in nature, but the wording is still based on an extractive fantasy, with clear rhetorical allegiances to colonialism and mining.

The hero's journey is an omnipresent cliché in Disney movies, video games, and the pop-cultural mindset. And perhaps certain old-school writing workshops. As a writer and a measured fan of Carl Jung, though, I'm not sure I can ever entirely escape the drive to weave certain archetypal elements into my creative work.

Yet in reality, the only "treasure" I returned with after that summer was moderate PTSD. Flashbacks, recurring nightmares about helicopters, an exaggerated startle response. And an acute bout of depression, especially when I decided to stop taking imipramine.

Shortly thereafter I gave up my obsessive drive to become an Outward Bound instructor, deciding instead to pursue what became my true callings.

Back in Boulder I started making more time for writing, playing my drum kit, drawing, taking photographs.

Becoming a professional writer is another path fraught with intimidating obstacles and pitfalls. Vows of poverty; bouts of terror. Hanging off financial cliff faces. And plenty of opportunities for your soul to get crushed. But on the whole, it's a gentler path, and one that suits me better than anything else—I'm quite content sitting alone in a room for hours, moving words around on a page.

I also started skateboarding again, after a few years' hiatus. I gave it up in high school to play football—something I thought my father wanted me to do. I was hopeless at team sports; mostly I sat on the bench, praying the coach wouldn't put me in. That was another year when I lost myself—a year that led up to my brief hospitalization for the *Hogan's Heroes*–induced panic attack.

Skateboarding gave me an outlet for my obsessiveness. Along with time, the exercise and joyful flow states I experienced while skating helped me gradually recover from the PTSD.

Though skateboarding in the late '90s was definitely not a space free from performative and often virulent masculinity. Once, in a minor argument, another skateboarder spitefully called me a *faggot*. Brian Anderson, the celebrated professional skateboarder, didn't feel comfortable coming out as gay until 2019—some thirty-odd years into his career.

And in some ways skateboarding was another way to abuse my body, to continue the obsessive grasping for hardness.

But decades later it's a pursuit I still love, and one that makes me feel alive, instead of like I'm about to die—at least *most* of the time. With age I've transformed my skateboarding from deep-end brinksmanship into lower-stakes, freestyle-influenced circus tricks and old-school basics; slappies and wallrides give me as much joy at fifty as they did at fifteen.

The skateboard community itself has transformed too; BIPOC, queer, and trans skaters continue to carve out space for themselves. Members of the queer community own and operate my local indoor park, Stronger Skatepark. Attending the monthly Queer Skate Night at Stronger feels like a revelation. It's a place where Raewyn Connell's belief in the existence of many, many masculinities—and many possible gender expressions—plays out in the most dynamic and kinesthetic of ways.

After coming out, Brian Anderson switched sponsors and now rides for one of my favorite skate companies. Not long after returning home from the Alejandro debacle in the late '90s, I purchased one of their decks.

The company's name is Antihero.

15.

I HAD ONE LAST ENCOUNTER WITH ROBERT BLY, IN his ponderously titled book *The Sibling Society: An Impassioned Call for the Rediscovery of Adulthood*. For whatever reason, I shelled out for the hard-copy edition, in which Bly rehashes his tired, morally panicked position that Western culture has gone astray because we no longer have father figures, that anything but a vertically organized, patriarchal society is doomed to failure.

"It's the worst of times, it's the best of times," he gripes in the book's insipid, practically mimeographed opening. "People don't bother to grow up," he goes on to write, "and we are all fish swimming in a tank of half-adults."

In one of his more unintentionally hilarious, *okay boomer* passages, Bly ponders, "How did we move from the optimistic, companionable, food-passing youngsters on that field in Woodstock to the self-doubting, dark-hearted, turned-in, death-praising, wised-up deconstructionist audience that now attends a grunge music concert?"

The Sibling Society attempts to resuscitate the old biblical notion of "putting away childish things" to become fully mature and responsible adults.

I felt sure Bly would view my skateboard as the quintessential "childish thing"—the simple little machine that helped me recover from an acute bout with PTSD, and that I used, along with writing and art, to reassemble myself after a disastrous summer, without the aid of any sanctimonious male mentor figures poisoning my beliefs.

In the end, though, I did adhere to this old biblical edict.

After giving up halfway through, I tossed *The Sibling Society* under my bed, where it remained.

Part Three

BARELY 7:00 A.M. AND ALREADY THE TROPICAL SUN heat-lamps our necks and calves. The scent of night-blooming ylang-ylang still perfumes the air; howler monkeys perform their horror-movie vocalizations off in the distance, many octaves below the morning's noise-rock concert of birdsong. At the tail end of a bird-watching trek in a Central American rainforest, our small group stands around chatting with our guide, Enrique Araya, a youthful-looking man in his early sixties with a compact frame and salt-and-pepper hair. A kind, animated smile. He grew up on this Pacific peninsula, deep in jungle terrain that's now a designated national park known as *Corcovado*—one of the wildest and most biologically intense places on the planet. His family tended several hundred hectares of rainforest, he tells us; they raised pigs, chickens, rice, and beans. The nearest populated town was a two-day walk. He and his father or siblings made the journey to purchase sacks of coffee and sugar, then turned around and hauled it all home for another two days.

So we're intrigued—especially our group member from Minnesota—when Enrique casually mentions that he spent six months in Minneapolis during the late 1980s, where he completed an English immersion program.

But just as he's relating to us how he survived an infamously cold Minnesota winter, he trails off midsentence, raises his hand, and says, "Listen."

I help shush the group into silence, then watch as Enrique cocks his head, turns his ear slowly eastward, back toward the grove of almond trees where ten minutes earlier we'd spotted yellow-throated toucans. Just after the toucans we'd lucked upon an Amazonian kingfisher perched on a barbed-wire fence, in its beak a silver fish that flashed so brilliantly in the sunlight that we first mistook it for a scrap of polished metal. Thanks to Enrique's keen senses, we've seen nearly fifty species of birds along this one-mile stretch of dirt road: blue-gray tanagers, black-headed trogons, scarlet macaws, yellow-crowned euphonias, swallow-tailed kites, red-legged honeycreepers—all the names of which I dutifully record and keep a running tally of in my marbled composition book.

I continue watching Enrique now as he further attunes to the jungle's cacophony. All I can hear are the *klink klink klink*s of orange-chinned parakeets, like someone plinking piano wires with an icicle. Among my favorite forms of birdsong, their calls always herald our arrival on the peninsula, wake us at dawn (along with the howler monkeys), add some chimey syncopation to the rhythm of our days.

Enrique, though, is on to something else—some other frequency for which my untrained ears have no bandwidth.

In *Horizon*, Barry Lopez observes that many Indigenous people seem to "pay more attention to *patterns* in what they encounter than isolated objects." Enrique's ability to distinguish certain faint calls among the riot of birdsong—and over our group's boisterous conversation—strikes me as remarkable in precisely this way. He politely urges our group to quiet down again, then points up to the sky, three or four hundred yards away.

A distance so great that, even through binoculars, we see nothing at first.

We continue following his directions and then, to our astonishment, witness a radiant bird materialize in flight above the distant canopy. It's a fireworks show finale of a specimen, with a large and brilliant red-orange beak, a two-tone yellow and vermillion breast, bright green legs. Only after we soak in the bird's blazing coloration and flight pattern does Enrique name it for us: a fiery-billed aracari. A rare variety of toucan that breeds only in this area; without Enrique's finely tuned ear and intimacy with this place's rhythms, we'd have missed it.

*

As a guide and friend for a dozen or more of our trips to this particular Central American rainforest—a place that feels more and more like a home away from home to my partner Lisa Mae and me—Enrique's helped us witness countless animals, birds, botanicals. All these years I've never seen him in anything but shorts, a T-shirt, and chukka boots, with a machete in a tooled leather sheath hanging from his belt—his standard attire as a wilderness guide, and a welcome stylistic departure from other local guides and gringo visitors who deck themselves out in nylon adventure-wear with zip-off pant legs and Panama hats. Something always bothers me about these other white tourists, with their multi-pocketed mesh vests and telescopic camera lenses. They all appear tainted with a similar strain of acquisitive desire—they seem to want to get in, score the requisite macaw and toucan photographs, then have locals chauffeur them off to $700-a-night ecolodges. But am I really all that different, truthfully, with my obsessive recording and counting of species in my notebook? Checking the boxes in the index of my old Golden Guide bird book enthralled me as an eleven-year-old, but there's always room for evolution in my avian preoccupations and my tendency to approach birds with the mindset of a certified public accountant. Could I give more credence, as Enrique does, to my senses—to the actual encounter with other beings—over my impetus toward naming and cataloging? If we happen to stumble across paradise, will our desire to analyze, acquire, and claim always spoil it for ourselves and others?

*

Sixteenth-century Spanish colonists named this land Costa Rica, meaning "rich coast," based on their false hopes of finding gold. Centuries later Jacques Cousteau apparently explored the nearby Golfo Dulce and declared it a "tropical fjord." The position of the so-called fjord, which is actually a gulf, leads me to believe that Project Plowshare very likely considered it as a site for a channel wider and more efficient than the Panama Canal, dredged via nuclear blasts.

For my own purposes, the gulf's perpendicular alignment with the open Pacific makes for an ideal surf break. Waves bend elegantly around the point and peel into a wide bay, where, if you time it right, you might trim down the shimmering blue face of the longest right-hander of your life.

The birding here is world-class, but would we return year after year if not for these waves?

On the exploitation scale, one could argue that surf tourism scores fairly low. The international and in-country flights do burn significant amounts of fossil fuels. Lisa Mae and I scoff at wealthy visitors who can afford the high-end eco-resorts, compared with our $70-a-night, open-air, thatched-roof *palapa*, but it's relative—the concept of leisure travel likely seems foreign to many local workers and ranch hands we pass on the dirt road. And surfing can bring out my territorial side—a quality of which I'm definitely not proud. Once, at a break we've surfed in near-private for almost a decade, a boat from the opposite side of the gulf motored up and deposited fifteen or more beginners in the water, pretty well killing the vibe. They were all American or European, but their guide was a local Costa Rican. Our German friend Eddy, the owner of the adjacent beachfront, had never seen anything like it, not since purchasing the property in the 1990s. Whose waves were these? Despite my dismay at the interlopers, they certainly don't belong to us. Nor to Eddy—in this country, one can own beachfront property but, fortunately, not the beach itself. If anything, the *tico* guide had a claim. His gaggle of mostly white tourists, who mobbed the break with little respect for us or the general etiquette of surfing? Not so much. If anything, they're a herald of things to come, just as we heralded changes for the locals during our first trip in 2008.

*

To make extra money as a young man, Enrique Araya panned for gold in the Rio Tigre. In the prior decades, American and Canadian mining companies ravaged the area, but in their wake the majority of small-time miners, or *oreros*, as they came to be called, were Indigenous people like Enrique, and there were only a handful of them.

As a young orero, Enrique spent long days in a snorkeling mask, plunging his face into a sluice box, sifting around for that infamous yellow glint. He once found 260 grams in an hour—the equivalent today of around $1,500 to $3,000 worth of gold. He tells me this with a gleam in his eye, gesturing all the while with his right hand, from which he's missing his index finger. Did he lose it while mining? If so, he never brings it up, and I never ask. What he does share is that, while he made a lot of money as an orero, it was a thorny chapter in his life. "I wasn't a good person then," he says.

*

In another nearby port town, the Dole corporation operated a banana plantation that remained productive throughout the 1970s. One day in the early '80s, Dole abruptly shuttered the plant, economically devastating nearly a thousand workers and their families. With limited options, the men flooded to the jungle by the truckload. What began with just a few families panning for flakes transformed, almost overnight, into a gold rush on par with any in North America in terms of its fervor, lawlessness, and violence.

On another excursion with Enrique, a taxi drove us through areas known as Dos Brazos and Los Tardes. They appeared to us like sleepy hamlets, but Enrique explained how dangerous they were during the gold rush era.

"You wouldn't want to come here during that time," he told us.

"Even if we were with you?" I asked.

Enrique shook his head, smiling in a way that I took to mean *especially if you were with me*.

*

While small bands of oreros still operate outside the law in Corcovado, most residents and tourists consider the Osa Peninsula a generally safe place. Yet certain hazards exist. Someone from the nearby town once crept into our lodging, stole a few of our friends' suitcases, dragged them off into the night. A local acquaintance found the ransacked luggage miles down the beach the next morning. Our lodge-owner friends grapple with these security issues, but prefer not to live behind security gates and barred windows, as most residents of the country's more populous cities do.

More concerning than the petty crime, though, is poaching. Rumors circulate about people smuggling animals, like the rare and endangered ocelot, out of the Park, as locals refer to it. They also take macaws—another threatened species—from their nests and sell them on the black market for a few hundred bucks.

During our first few visits in the late 2000s, we interacted briefly with a Canadian immigrant who crafted her own artisanal chocolate from local cacao. She sold it by the bar at the Friday evening farmers market held at a local open-air tavern. People coveted her chocolate—they queued up for first dibs and she often sold out before the night's end. She lived much closer to Corcovado than anyone else we knew, way up north in the dense jungle, where she gained a reputation as a fierce defender of wildlife. She owned a BB gun and wasn't above chasing off poachers in the middle of the night. Based on what Enrique told us about the area's history—and the fact that certain oreros still linger, often moonlighting as poachers—her actions seemed both heroic and reckless. Was she an eco-warrior willing to sacrifice her own safety for a higher cause? In many regards, yes. Woefully underfunded and understaffed, the government employs only a handful of rangers to patrol the Park's four hundred square kilometers. And from a gendered perspective, she was beyond gutsy in confronting perpetrators who we can assume were male. Given the Dole plantation's sordid history of exploiting workers who eventually turned to mining and poaching, one might also place her on the spectrum of white saviorship, seemingly unsympathetic as she was to the plight of generational oppression and poverty, and the extremes to which these forces can drive people.

Regardless of how you view her actions, we were shocked and dismayed to learn, during our fourth or fifth visit, that poachers had dragged her from her home late one night and took her life in a gruesome execution-style murder. Some locals seem to have a sense of who perpetrated the act, but feel too afraid to utter any names. Her case remains unsolved.

*

As far as I know, the Bible never mentions the equatorial tropics. Why, then, do Western cultures so often equate the tropics with the concept of paradise? We might trace the phenomenon back to the so-called discovery, by French and British imperial fleets, of equatorial locales like Tahiti as recently as the eighteenth and nineteenth centuries—the fantastical descriptions of which took hold of their countrymen back home. European artists such as de Stuers, and, later, Henri Rousseau, began producing romanticized, paradisial landscapes that always exempted the violence and exploitation of labor and the natural world inherent to the project of colonization. Created in the mid-seventeenth century, Flemish artist Jan Brueghel the Younger's painting *Paradise* foregrounds a scarlet macaw perched on a tree above a placid lion—species that, in reality, exist on entirely different continents. In the tableau a variety of predators, including what looks like a domesticated house cat, lounge on an Edenic meadow, co-existing peacefully with commonly preyed-upon animals like shorebirds, sheep, and an antelope-like creature. We might ascribe these fanciful juxtapositions to Brueghel's artistic sense of whimsy, or to biblical notions of the "lion lying down with the lamb." Though from a contemporary perspective it's difficult to ignore the simultaneity of his artwork with, say, the Dutch East India Company's brutal conquests of people and land. Does the painting represent an earthly Shangri-La, or a kind of sanitized, imperialist cabinet of curiosities?

Can "paradise" only ever exist in the imagination? If so, what critical omissions or substitutions perpetuate the fantasy?

*

Enrique Araya was fortunate to escape the hardscrabble life of an orero via the burgeoning tourism industry. In the 1980s he worked at a lodge near an encampment named Sirena, not far from where he grew up. He took care of various tasks around the property, until one night the lodge owner approached him with a panicked expression. The owner explained that a young Canadian couple failed to return from an afternoon hiking excursion.

He entreated Enrique for help.

"Give me a liter of water and a flashlight," Enrique responded.

*

During a trek near Los Tardes, Enrique hiked with us along a rutted jeep road, up to a grassy meadow, where we spotted a primitive entry sign to Corcovado. The second we stepped out of the meadow and behind the rainforest's dense curtain, the temperature and humidity mushroomed. The light dimmed and the atmosphere grew steamy; the volume of cicada buzz and monkey calls intensified. The jeep road, I realized, demarcated a striking ecotone, allowing one to stand with a foot planted in two different climates. As we continued the trek, Enrique warned us not to pause for too long. We soon learned why: any stillness gave fire ants the chance to stream up your boots and aggressively bite any exposed skin. The trail itself was muddy and steep. Enrique also advised us not to stray from it, for fear of deadly poisonous fer-de-lance snakes.

Corcovado National Park comprises over 160 square miles and some thirteen ecosystems. That morning we skirted only a tiny portion of what I can honestly say was among the most formidable terrain I've ever experienced. Those who brave it certainly encounter wonders: at the end of our hike, Enrique trained his spotting scope on a rare white-tipped sicklebill hummingbird perched near a small waterfall. She allowed us to marvel at her curiously arched bill and her purple and green plumage for several breathless minutes.

I'd like to tell you I've seen much more of Corcovado, but we only occasionally deviate from a one-mile stretch of surfing beach and nearby forest during our visits. In my early twenties I thought I wanted to spend every waking moment in the deep backcountry, exerting myself on trails and rock walls. By my thirties and forties, I grew to prefer the privilege of spending my yearly vacation tooling around in the surf on a nine-foot longboard, then lounging on the beach with friends or wandering around in flip-flops, admiring birds. Spending so much time in the ocean, living in accord with the shifting tides, makes for my most memorable and blissful encounters with nature. I still love hiking and the occasional short backpacking trips, but I just no longer feel the urge to conquer anything. Especially Corcovado, for which I maintain a deep respect and healthy distance.

*

Many tourists who visit the peninsula, on the other hand, do feel inevitably drawn toward Corcovado—including the aforementioned young couple from Canada. Their fateful excursion took place in the days before it was forbidden to enter the Park without a designated guide. After accepting his boss's plea for help, Enrique, not yet a guide in any official capacity, ventured into the jungle at around 5:30 or 6:00 p.m., at which point the equatorial darkness had already set in.

He tracked the couple for more than seven hours, until, at around 1:30 a.m., he found them huddled in the pitch dark, vibrating with fear—not unlike Alejandro in his cave-like shelter up in the San Juans. The couple was lucky: this is the same jungle into which seasoned Alaskan outdoorsman Cody Roman Dial attempted a brief solo excursion in 2014 and was never seen alive again. Cody's father, a well-known Alaskan trekker and *National Geographic* correspondent named Roman Dial, organized intensive searches and even hired former military personnel to investigate. Even so, Cody's body wasn't recovered until two years later, hidden as it was beneath a fallen tree.

The next morning, Enrique personally delivered the Canadian couple to the airport. As passengers boarded the cramped twin-prop plane, Enrique noticed that it seemed overloaded with luggage. He suggested to the crew that they'd taken on too much weight. When the pilot ignored him, Enrique urged the Canadian couple to wait for the next plane. Following his advice, they watched together from the ground as the first plane labored to gain enough speed for takeoff.

Instead of lifting off, the small craft pitched end over end into a swamp at the short runway's terminus.

Enrique sprinted down the tarmac to the inverted plane, where he found the pilot hanging upside down and purple-faced, strangled by his own seat belt. Enrique cut him loose with his machete and dragged him out the window. He also rescued a woman with a broken arm; she seemed relatively fine, but died some days later in an American hospital, possibly of an unseen head injury.

As for the Canadian couple, they watched in awe as the man who saved their lives twice in twenty-four hours pulled out those buried deeper in the plane before it burst into flames. They eventually made it home safe to Canada, where they commissioned for him a painting of a red-capped manakin—Enrique's favorite bird.

*

Once, after another excursion near the Rio Tigre, Enrique invited us to his house for homemade fruit juice and slices of watermelon. On the wall near the kitchen, above the hammock, five or more government-issued plaques commemorate his rescue efforts. Near the plaques hang framed magazine covers with colorful photos of various wildlife. He explained how he spent six years helping a well-known photographer track and document a puma and its family in Corcovado. He wouldn't say it then, but I know from internet searches and my own experiences that Enrique's considered one of the foremost wildlife trackers in Central America, if not the world. He also shared photos of his girlfriend and children. His three daughters and grandchildren all live in or near the city, and, as Enrique shared with good-humored resignation, none are particularly interested in the natural world.

Earlier that day, after trudging up an exceptionally muddy jeep road above the Rio Tigre, Enrique showed us the black fungus devastating local cacao crops. In a small grove of trees he found only one or two viable cacao fruits; he knifed one open and shared the purple, fruit-covered seeds with us. While we gnawed semisweet flesh from the seeds, he ate his right off the machete blade. It tasted nothing like chocolate; it had more of a musty, citrus tang.

"Things are strange and unbalanced," he told us. "It's not just the cacao. Other types of trees fruit twice in the same year instead of just in spring. Birds migrate in and out at unusual times. The weather's getting hotter, wetter, more unpredictable. I can't even grow beans or rice in my little garden patch anymore."

It was disheartening to learn that, for those with heightened pattern detection, this place feels threatened by climate change, despite its outwardly lush and teeming veneer.

It was July and hundreds of thousands of acres of forest were burning back in the States. The entire West mired in continual drought, the Colorado River visibly drying up, to everyone's horror. Compounded by deep racial divisions in our country and the frightening rise of neofascism, things felt dire.

The evening beforehand Lisa Mae and I celebrated our wedding anniversary over dinner, during which she wondered out loud how much time I spend considering the end of the world. In a teasing response, I encouraged her to expand her reading list beyond all the postapocalyptic novels she'd been devouring. Though truthfully I knew exactly what she meant. It feels like so much hangs in critical balance—American democracy, the climate, arctic ice, the fate of so many animal species. The sense of the entire human project reaching an endpoint can feel exacerbated for those of us without what author Sarah Manguso describes as the "ongoingness" that people with children often experience. Given the state of things, we talked about relocating to a place like Costa Rica. The idea compels me, to some degree—at one point on this trip Enrique even asked us if, after spending so much time here, we might consider buying property? Though I personally feel conflicted about entering (or reentering) the long tradition of upwardly mobile white folks relocating to already-inhabited places to escape troubles in their own countries of origin. The northern regions of Costa Rica already seem elbow to elbow with American and European immigrants. Not to mention the regular encounters with poisonous spiders, or the way tropical termites constantly drill through the roof beams, coring out thick dust that coats everything below—bedsheets, pillows, your toothbrush.

I honestly always feel relieved to land back in home in Oregon—to our modest garden and our bug-free bed. To careers we love and a web of connections with dear friends. And with the forests, volcanic mountains, and beaches of the Pacific Northwest. It's painful to even imagine leaving our own flawed paradise.

*

Later that day, while admiring Enrique's commemorative plaques, an inevitable question percolates into my consciousness: If we've lost our way—*we* meaning Americans and capitalist societies in general—as hopelessly as those Canadians in the jungle, what guiding answers or directions might Enrique have for us?

I hold my tongue, for fear that my inquiry might place Enrique in an uncomfortable position. The fantasy of being rescued (or healed, or spiritualized) by Indigenous people is equally pernicious as the ideology of white saviorism. An extractive form of hero worship, perhaps.

Though would it be wrong to attempt, in our imperfect ways, to learn from and emulate Enrique and others like him, not as heroes but as wise elders? As Robin Wall Kimmerer writes in *Braiding Sweetgrass*, our future as a species relies on more and more of us becoming naturalized to the places we live—to enter into deeper, more reciprocal relationships with the plants, the soil, the air that sustains us. Equally compelling is the notion, proposed by authors David Treuer, Roxanne Dunbar-Ortiz, and others, of returning the stewardship of America's national park lands to their original Native American inhabitants—so many of whom were defrauded and evicted during the creation of "America's best idea."

Enrique is highly respected and even coveted as a Central American guide. He seems to know nearly everyone who lives in the vicinity of the Park; most treat him with admiration and reverence. Though once, when speaking with a younger guide, I thought I might have detected a slight tone of condescension toward him, as a former orero. And who knows what harm he may have caused during his younger, rougher years? Still, Enrique's history—his having grown up deep in the jungle, then shifted course, as a young man, into a miner focused on money, acquisition, power—*you wouldn't want to know me then*—to an individual now dedicated to protecting and educating others about the same lands he once mined—contains a narrative arc that I think we might further cultivate in our own personal and collective stories and actions during this era of late-stage capitalism.

*

In the summer of 2021, on a trip to visit my parents and survey Colorado's National Mining Hall of Fame and Museum for research purposes, I encountered what I originally considered a salient counterpoint to Enrique's life narrative.

On a warm, sun-drenched July afternoon, my mother, stepfather, and I biked the twelve-mile Mineral Belt loop above Leadville. The town itself sits at 10,200 feet—the highest occupied city in the United States. The Mineral Belt Trail led us up another 750 feet in elevation, where we encountered striking contrasts: stunning fourteen-thousand-foot peaks, aspen groves, and conifer forests background the ravaged landscapes of Fryer Hill and Oro City, where gold and silver rush–era miners wounded and sluiced and dynamited the earth into oblivion, leaving in their wake what presents as something akin to a WWII battlefield. Around every corner sit derelict timber structures still stinking of creosote. Massive slag heaps rainbowed with toxins and tailings ponds the color of cherry cough syrup. Now designated as the California Gulch Superfund Site, ongoing cleanup efforts attempt to mitigate the continual leaching of arsenic, cadmium, copper, and lead into the soil, groundwater, and food chain.

I'd never seen anything quite like it, even as a born-and-bred Coloradoan.

After pedaling up the long, winding hill, only to cruise at top speeds down another alongside my beaming parents, all of us enjoying the scenery and sun, I experienced the familiar if ephemeral feeling of joy even amid the cataclysm.

After the descent, and on a whim, we followed a signpost leading us off the main bike path to a historical site maintained as an outpost of the National Mining Hall of Fame and Museum. Just by accident, we landed upon the storied Matchless Mine. Among the most prosperous silver mines in Colorado during the mid to late 1800s, it profited its owner, H.A.W. Tabor, as much as $80,000 a month—a dizzying sum by turn-of-the-century standards. He and his second wife, Elizabeth "Baby Doe" Tabor, lived notoriously opulent lifestyles and were often referred to as the Silver King and Queen of Colorado.

Now, all that's left of the Matchless are seven gaping pits and a dilapidated wooden toolshed perched above them.

I'd heard the name Baby Doe Tabor countless times during my Colorado childhood. A chain of mining-themed restaurants bore her nickname. An approximation of her ornate bedroom once existed in the Glenwood Springs Frontier Historical Museum—the place I visited during grade school field trips. From these references I knew of her only as the wealthy socialite wife of a mining tycoon.

Now, reading interpretive signs posted around the Matchless Mine and talking with on-site historian Christine Whittington, I was astonished to learn that the former Silver Queen lived out the last thirty-some years of her existence inside the aforementioned toolshed. An impoverished widow, hovering on the edge of the abyssal mineworks, she apparently clung to the belief that the mine might once again produce silver.

I hadn't even set foot in the actual mining museum, and already I felt I'd struck dramatic gold in the form of a cautionary tale.

*

Elizabeth "Baby Doe" Tabor was a woman whose operatic life's narrative—an actual opera was written about her, plus many biographies and a film starring Edward G. Robinson—played out just over an hour from where I grew up. She was born into a middle-class Irish-Catholic family that lost nearly everything during the 1874 fires in Oshkosh, Wisconsin. She eventually married into the more established family of Harvey Doe, who brought his new wife along to seek fortune in a Colorado gold mine owned by his father. As one of few women in the town of Central City, the local men apparently worshipped her as a kind of classically beautiful goddess. To assist her husband, she dressed in men's work clothes and helped in the mines, causing a scandal among Colorado's elite women.

After growing disillusioned with Harvey Doe's drinking and business failings, Elizabeth began an affair with H.A.W. Tabor, a married man who, thanks to his stakes in various silver mines, was among the richest in the country. They made headlines—and the lips of every gossiping Coloradoan—by falling madly in love and divorcing their spouses to be with each other. They lived the most lavish lifestyle imaginable, in mansions and estates and hotels around the country. Their private suites in Denver's Windsor Hotel were rumored to have mirrors made of diamond dust and a gold-leaf bathtub. In a tacky, nouveau riche move, they named their second daughter Silver Dollar Tabor. Despite his ambition, Horace Tabor was once known as a generous man, often willing to help out his fellow prospectors. After striking it rich he built opera houses and funded civic works; he served brief terms as a U.S. senator and lieutenant governor of Colorado. Yet, while the Matchless Mine raked in millions, his laborers worked for the average rate of $2.50 a day, seven days a week, with no medical care or health insurance. Tabor operated his own private militia, ostensibly for protection, but largely to keep his oppressed workers in line and violently suppress any strikes or union-organizing efforts.

Elizabeth was very often maligned by the press and upper-crust Denver socialites, who didn't approve of her divorce and scandalous marriage to the senior Tabor. Her beauty and brash sexuality offended them; they considered these assets as tools in her scheme to acquire Tabor's money. Whereas I find it hard not to admire her verve in pursuing what she desired, and in jettisoning the social mores of her time—especially the tightly circumscribed codes for feminine behavior and gendered clothing choices. As the Silver Queen she also championed the women's suffrage movement, providing feminist activist groups of the era with generous donations and free office space.

*

Like Enrique Araya, Elizabeth Tabor spent her formative years deeply embedded in the mining industry. Similar to the younger Araya, she was profoundly materialistic in her young adulthood. And, like Enrique, she lived to see dramatic reversals in her destiny.

*

During the Panic of 1893, the price of silver collapsed. Subsequently the Tabors' fortunes evaporated in a matter of weeks; the majority of the family's holdings were sold off to cover outstanding debts. Horace Tabor attempted last-ditch investment schemes in various parts of the world, including mining ventures in the San Juan range. A century later, during my Outward Bound and Wayfinding days, I likely hiked near arsenic-laden slag heaps left by his desperate gambits.

When his schemes failed, the former U.S. senator and Silver King once worth nine million dollars returned to Leadville as a sixty-five-year-old day laborer with a pickax and shovel. Based on what remained of his reputation, he was eventually given a handout job with the Postal Service in Denver. He died of appendicitis in 1899. According to apocryphal and quite dubious lore, on his deathbed he implored Elizabeth to hold on to the Matchless Mine and never let go, in case the price of silver came back.

Whatever her husband did or did not say on his deathbed, Elizabeth Tabor clung to the Matchless Mine. She eventually relocated from Denver to inhabit the former toolshed above one of the Matchless's seven now-flooded main shafts. She brought her children in tow, but they both gradually abandoned her. The conventional narrative states that she dwelled there in poverty, semi-squalor, and madness, surrounded by hoarded newspapers, for thirty years, selling off her jewelry while her brother made secret payments to a grocer to keep her fed. Many citizens of Leadville viewed her as a crank. She reportedly roamed the town's streets at night, dressed partly in men's work clothes, with newspapers stuffed in her jacket and burlap sacks wrapped around her feet for protection from the elements. She greeted any uninvited visitors at the shack with harsh words and the sight of her rifle. Some visitors to the area near her cabin claimed to have heard her giving orders to an imaginary maid. Many viewed her Icarus-like downfall as punishment for her former ostentations and what they judged as the immodest home-wrecking behavior of her youth.

Elizabeth survived not only her husband, but also her cherished daughter, Silver Dollar, who was eventually found scalded to death in a Chicago boardinghouse after trying her hand at writing and dancing; suspicion fell largely upon her gangster boyfriend. Elizabeth was forever gutted by this loss, until her own death by apparent heart attack in 1935. Her frozen body wasn't found until several days after her passing. The *Denver Post* headline on Friday, March 8, 1935, read, BABY DOE TABOR FREEZES TO DEATH WHILE ON GUARD AT MATCHLESS MINE.

Elizabeth Tabor's story contains all the trappings of mythic tragedy. It initially captivated me as such, and as a pointed contrast to Enrique Araya, who ostensibly cast off his enthrallment to an extractive narrative in ways that Tabor never could. But I now view my own initial judgments—and my hasty comparison of her life's narrative to Enrique's—as itself a kind of exploitative fantasy.

After reading more about her, I've come to admire the raw grit it took to live for thirty years in a cabin at eleven thousand feet with no electricity or running water, where she would've encountered some of the harshest winter-weather conditions possible in the lower forty-eight states, nursing her self-described mental anguish with sparse access to professional health care. Why do we lionize Henry David Thoreau for spending several seasons in a cabin an easy twenty-minute walk from his mother's house, while hardier souls like Elizabeth Tabor are remembered only for their youthful beauty?

I can also certainly empathize with the desire to hole one's self up in the wake of loss. And I know something about living in (or at least visiting) remote places when my own mental health was in a desperate state.

*

In late 2015, I made some medication changes and several consequential career decisions in short succession.

Desiring more time for my own writing, I decided to step down as director of a beloved community arts nonprofit—the Independent Publishing Resource Center, or IPRC—after almost a decade of service. To support myself, I began teaching part-time in a new low-residency MFA program based in eastern Oregon. I also spent weeks applying for a tenured professorship in a much more established university writing program back in Colorado—though, at the last minute, I made an impulsive decision to remove myself from consideration. Generated by the retirement of my former writing professor and mentor who coached me through the early job application stages, the Colorado teaching position seemed like an auspicious match. But at the time, my partner and I were ambivalent about leaving Portland, and her acupuncture practice would be nearly impossible to rebuild in another state.

Not long after, the eastern Oregon–based low-residency MFA program ceased operations. Now, instead of potentially enjoying a permanent and gainful university teaching position, I was essentially unemployed. Around the same time, the Independent Publishing Resource Center lost its physical space and came close to shutting down.

Before dawn one November morning, severe depression and anxiety hit me with the force of a viral influenza. My limbs trembled uncontrollably; an icy sweat broke across my chest and neck. It was the last night of sleep I'd have for two or more weeks, my mind so racked by intrusive thoughts and obsessive regret about leaving the IPRC and ruining my once-promising career.

Within a matter of days I lost five or more pounds; I was so emaciated I couldn't swim in the heated pool at the gym without bursting into fits of shivering. Friends sometimes failed to recognize me; my partner described my constant expression as a depression mask.

This poisonous fog enveloping my emotional life terrified me and my family. It was the holidays, and I tried to seek professional help to sort out the medication issues, but I had little luck.

Every waking moment of December—and the months to come—felt nearly unbearable. Trapped under the mental slag heap of *what could have been*, I lost the ability to imagine a viable future for myself.

Instead I perseverated on the missed opportunity for a triumphant homecoming back in Colorado—to make what Joseph Campbell might have classified as "the hero's return," even though I'd long ago abandoned my attraction to Campbell's work, at least outwardly. My Colorado-based mother's eventual diagnosis of triple-negative breast cancer only exacerbated the feeling that I should've moved back to my home state when I had the chance.

Another element of my despair was my partner's and my decision not to have children. Elizabeth Tabor actually lost her younger daughter, Silver. There's absolutely no comparison to losing a flesh-and-blood child, of course. But during the depression I grieved the daughter I never had—a peculiar kind of loss similar to what sparked author Andrew Solomon's long bout with melancholy, as described in *The Noonday Demon: An Atlas of Depression*. One sleepless night on the couch, I felt her phantom presence there in the room with me, her confusion and sadness about her own absence hanging thick in the air.

As Tabor wrote in a diary entry dated 1922, "I dreamed of being with my darling Child Silver. She was in a sort of white all snowy white cheap gown & She looked pale & white & weak & was worried in trouble . . ."

I also fantasized about exiting my life and holing up somewhere remote enough that no one could witness what I perceived as the failure I'd made of myself. On Christmas Eve I drove aimlessly around the remote outskirts of Portland, seeking out some small parcel of land on which I might build a modest shack and disappear.

As the illness progressed, the idea of hospitalization loomed as an ominous possibility, but we had a trip booked to Costa Rica in early January. Twenty or more people were relying on my partner to lead a retreat they'd already paid for. She had to go, and there was no way she was leaving me home alone.

In Central America, the cicada buzz echoed the state of my nervous system. I woke up hours before dawn with a feeling of low-voltage electricity zapping my heart muscle. During daylight hours I recoiled from the birdshriek and the wind-rattled palm fronds slicing down at my wrists.

One thing I could do was surf.

Out in the water, though, I obsessed about paddling off into the gulf, just as, during a previous depression in New York City, I strayed way beyond the break during solo night surfs at Rockaway Beach.

When even surfing brought me no solace, I used the faint Wi-Fi signal to search the term for a stark inability to enjoy once-pleasurable activities: *anhedonia*.

One brief, bright spot of the trip occurred during a group outing with Enrique Araya. Following his acute sense of hearing, Enrique scrambled off the road into a ravine, then quietly invited everyone to join him. One by one we dropped into the gulley to view an orange-capped manakin perched deep in the brush, its head like a perfect scoop of sorbet.

But in general the place where I spent the most blissful weeks of my life now felt like the opposite of paradise, in a way that perhaps echoes the source of Elizabeth Tabor's previous wealth becoming the site of her thirty-year penitence.

*

Another aspect that changed my mind about Elizabeth Tabor—and further expanded upon the tiny cabin, constructed from my growing affections for her, that now lives in my consciousness—was learning that her survival depended in part on expressive writing. During her semi-monastic years at the Matchless she produced hundreds of notes, diary entries, and various ephemera. She described the fragmented collection as her memoirs. Important pieces of these memoirs were likely lost when scavengers raided the cabin after her death. But many pages do remain in historical archives.

Tabor composed with a loose and rambling style, casting off rigidity and formal writing conventions so prevalent in the era. She employed poetic repetition and made exuberant use of ampersands to string along torrents of prose. By contemporary standards, we might describe her various literary artifacts as lyrical essays, prose poems, or fragmented memoirs that, despite their unedited, raw quality, often achieve mystical crescendos.

Just as remarkably, she drafted on scraps of paper and envelopes, not unlike Emily Dickinson, then later revised and organized these fragments into booklets that we might even label as prototypical DIY zines or chapbooks.

In *Baby Doe Tabor: The Madwoman in the Cabin*, author Judy Nolte Temple writes:

> Fragmented, jumbled, compromised, interrupted. These are the telltale signifiers in the history of women's writing . . . Emily Dickinson warned, "Tell all the Truth but tell it slant—" and closed Poem 1129 with this lesson: "The Truth must dazzle gradually / or every man be blind—" . . . Lizzie Tabor, however, wrote with blinding white-hot intensity in her effort to tell her truth, which survived only in fragments. In geological parlance, however, fragments are exceedingly interesting phenomenon. They indicate disruption is happening, that something is undergoing dissolution. How fitting, then, that this disruptive femme fatale left her legacy in fragments that disrupt the legacy of Baby Doe.

Thanks to Temple's painstaking research and transcription efforts, we have excerpts like these from Elizabeth's journal: "My love is like a shut-up fire in a tomb & my children hold the key. Extreme mental anguish."

During the most severe days and weeks of the depression, my own journal entries struck some similar tones. Many consisted of my naked pleas for rescue from a higher power that, at the time, I was doubtful even existed.

If there were a god, I wondered on the page, how could she allow me to feel so abjectly miserable?

Whereas I had the benefit of a loving partner, professional counseling, medication, acupuncture, a circle of friends and family, and, eventually, some unexpected career opportunities, Lizzie had only herself and her journals.

*

In mid-2016 I accepted a one-year visiting professor position at Evergreen State College in Olympia, Washington. I made the two-hour commute up the I-5 corridor in the hammering rain, bingeing on murder podcasts, imagining self-inflicted car crashes, then white-knuckled my way through lectures. It was a particularly contentious year at the college, especially after the 2016 presidential election; various faculty members waged internal warfare every day on a campus-wide email forum. The battle culminated with one white male professor slagging the school on Tucker Carlson's Fox News show; a couple days later a serious bomb threat by white supremacists sparked a campus-wide evacuation.

An Evergreen dean of hiring had enticed me into the visiting position with the chance to apply for a future permanent position, but in spring 2017 no one responded to my application with even a simple rejection form, despite my solid year of service. It was probably for the best, as the distance between Olympia, Washington, and Portland, Oregon—along with my persistently gloomy mood—severely strained my marriage.

A lifeline came in the form of an unexpected invitation to cover a memoir course for a writing professor in the process of retiring at Portland State University. Eventually the one class turned into a full course load and a series of temporary appointments.

Working with supportive colleagues and talented students at PSU lifted my spirits. As did an ongoing, daily text exchange with a close friend recovering from opiate addiction; we helped keep each other alive. Riding with friends at Stronger Skatepark helped regulate my moods. So did snowboarding with Lisa Mae on Mount Hood—our relationship more solid than ever after all we've lived through together. My mother recovered from cancer, thankfully, and despite losing the physical space we'd worked so hard to build out, the Independent Publishing Resource Center found a solid new location and fresh leadership, to my immense relief. After a twenty-year hiatus, I took up the drums again, and, with two close friends, formed an indie-rock band named Golden Tiles—a reference to the sublime, sun-glazed portal that so captivated me in Barcelona's Sagrada Familia cathedral.

I don't want to sugarcoat reality with these details from my slow recovery: the severe depression and daily suicidal ideation lasted a full eighteen months and nearly destroyed me and my marriage.

The concept of recovery wasn't even part of Elizabeth Tabor's lexicon. Yet she endured, and though it's been mostly ignored for decades, so does her writing.

*

In *Horizon*, Barry Lopez expresses his abiding interest in the act and utility of *dreaming*. During a chapter about his participation in an archeological dig in the remote arctic, he continually ponders what the ancient Thule people dreamed about. Especially during the long polar nights, when they likely slept for twelve-hour stretches or more, and are thought to have actively sought out a kind of liminal dreamworld. What was the content of their dreams, Lopez wonders, and how might their dreams have helped them survive, individually and as a society?

We don't always know the true contents of her mind, but surprisingly we do know much of what Elizabeth Tabor dreamed about. As part of her memoirs, she kept detailed diaries that she named Dreams and Visions.

In the recounting of one particular dream, she imagines riding in a carriage with her beloved daughter Silver, heading toward the Matchless Mine, where they encounter a miraculous sight:

> . . . to our right between the mountains appeared a great and mighty and wonderous rainbow O with such a wide part of it in Ruby glittered light all aflame in Ruby fire & the Ruby colored red made the sky Ruby red & O it was so bright it dazzled you to look at it O too glorious to look at & we saw other colors but the wonderful Ruby color aflame dazzled it all it sank between the mountains & at

no time could we see it all we rode along slowly &
all seemed peaceful & arrived home and I seemed to
be in bed with someone I think Miss Dr. Fantz . . .

It's remarkable that, in this reverie, we witness Elizabeth's evolution from the coveting of physical wealth—all the actual gemstones and rare metals she once owned—to what we may well interpret as a more natural and spiritual wealth—"glittered light all aflame in Ruby fire." Pliny the Elder might have noted that this dream's riches derive from above, from the sky, rather than from "the things that destroy us and drive us to the depths below."

What to make of the latter half of this ecstatic vision, wherein Elizabeth dreams herself into bed with a doctor's wife? Later in the same dream, her first husband, Harvey, lies down between them. The passage crackles with uninhibited sensuality in ways perhaps comparable to Walt Whitman, whose *Leaves of Grass* kept me company during my Outward Bound solo and contributed to my own lust-aching dreams up near eleven thousand feet. And with my contemporary, queer sensibilities, it's hard not to read into Elizabeth's breathlessly detailed description of a rainbow just before her marginally chaste ménage à trois.

*

I once witnessed Barry Lopez fumbling with a university classroom computer just before delivering a talk on environmental writing to dozens of eager conference attendees. I felt a little embarrassed for him, familiar as I am with the sense of ineptitude and shame that computer problems can spike within a teacher, especially in a room full of young, tech-savvy students. Lopez eventually had to call in an employee from the college's IT department, delaying the session start time by another ten or more minutes.

To my surprise, instead of requesting help cuing up a slide presentation, Lopez asked the employee to remove the computer from the room. Only after the tech lugged the machine out into the hallway did Lopez commence with his talk.

Though he always composed his personal letters to me on a manual typewriter, I have trouble believing Barry Lopez was a complete luddite. I doubt he generated manuscripts like his five-hundred-page opus, *Horizon*, entirely without the aid of a computer. But the symbolic gesture he made that afternoon stays with me, over and above anything that was spoken during his lecture. In an act of subtle performance art, Lopez communicated that, much the same way Thoreau witnessed the railroads "riding us," computers and technology inevitably transform our relationships to the world and each other, in ways that are difficult to detect from inside the culture.

The memory of Lopez's radical act of pedagogy returned to me during COVID-19 lockdowns when I spent mindless hours in front of the TV or doomscrolling social media. Twenty or more internet tabs remained forever open on two separate computers; I checked my three email accounts fifty or more times a day. I wrote and read almost nothing during long stretches of quarantine. Due to travel restrictions, we canceled trips to Costa Rica. Then I began noticing a recurring dream, in which a malfunctioning cell phone thwarts my urgent need to communicate with Lisa Mae or some other loved one—the screen possessed by a frenetic string of incongruent images, garbled advertisements, static. Nothing I do can fix or even shut down the phone. Is this my subconscious mind's attempt to purge all the unprocessed information? Whatever the case, the nightmare often dominated my sleep and left me feeling disappointed in the anxiety-provoking and frankly stupid content of my dreams. Why couldn't I dream of something more profound or meaningful? Why am I always so isolated in the dreamworld?

Once COVID restrictions eased, I attempted to concentrate again on my own stories. Just the act of reading words on the screen left me feeling exhausted. I sometimes fell asleep sitting at my computer, chin in hand. Fatigue, irritability, inability to focus—they felt especially acute when I shut off my phone and internet connection, and I came to understand them as withdrawal symptoms.

To counter the withdrawal, and inspired by Ross Gay's *The Book of Delights*, I began cultivating more of my own sensory-based pleasures. In the mornings I picked and deeply inhaled the scent from sprigs of lavender, thyme, basil, rosemary, and Corsican mint from our little garden. During a weeklong writing retreat at the Oregon coast, in a tiny trailer with only an analog record player for entertainment, I had what felt like a subtle but meaningful spiritual experience just by refocusing my attention on writing, the ocean, the feeling of a sea-cooled breeze on my bare calves. These simple practices helped me feel more alive and connected; no international travel was required. And while the cell phone nightmares haven't entirely left me, my dreams have expanded into somewhat more fertile territory again.

Beyond our own selves—in the realms of culture, race, politics, and environmentalism—how do we make more space for dreams and the act of dreaming? Especially in an era when we face so many apparent endings and extinctions, and when our literature tends so frequently toward the dystopian? We might look to contemporary speculative fiction writers and social justice activists like Walidah Imarisha and Alexis Pauline Gumbs, who believe that, over and above the fetishization of technology, science fiction allows us to dream up futures that currently seem impossible to us.

*

Remarkable as her own dreams and writing were, I don't think we should let Elizabeth Tabor off the hook for her deep imperfections. She and her second husband profited off the forced removal of Utes and the ravaging of the earth—scars of which still remain in Colorado's landscapes and waterways—and off the backs of exploited workers, many of whom likely hailed from my ancestral Cornwall. She relished her role as the symbolic queen of white American striving and greed. And throughout much of her thirty-year vigil she continued to harbor fantasies of restoring the Matchless Mine and her opulent lifestyle.

But didn't she, in the end, reclaim her life in creative and spiritual terms, in ways that resonate with Enrique Araya's own transformation? The eighty-year-old Elizabeth who occupied that one-room cabin interests me infinitely more than the youthful Baby Doe who lived in mansions and hotels. In her later writings I find myself attracted to the vulnerability she expresses from within the cracked container of her life. Though several Leadville residents who knew her claimed she was often quite content with her hermetic existence, and certainly not insane.

As the preteen Silver Dollar Tabor wrote in a letter to her aunt: "No one knows where momma and I are. We have a very quiet and secluded little spot and it has been the only place for momma right now and it is very lovely. It seems the Angels must have taken us to it so please dear Aunt Claud do not mention our names to anyone in Denver."

I'd like to believe Elizabeth's intermittent contentedness had something to do with her unalloyed attention to her natural surroundings. Who knows what she witnessed during those three decades in the shadow of Mount Massive and Horseshoe Mountain? Perhaps she befriended mating pairs of Wilson's warblers, or considered red foxes and aspens as her kinfolk. Did she appreciate the butterscotch scent of lodgepole pine bark in the summers? Or cherish the rare stand of bristlecone pines—known as the oldest and hardiest tree species on the planet—that still exist within a mile of her cabin, and that was pointed out to me by my mother, an amateur naturalist in her own right?

Tabor's turn toward religion was read by some as another sign of her madness. Her Christian zeal does appear overly rigid and even fanatical at times—we see this in her codependent, smothering relationship with Silver Dollar, whom she tried to control and prevent from sinning. This was certainly ironic, given Elizabeth's own lifestyle as a young woman, but also prescient, considering Silver Dollar's fatal plummet into addiction and enmeshment with unsavory men. Tabor's mysticism might also speak to us about the need for sustenance beyond what society and money and striving can provide—I certainly could've used some of her transcendent grit during my own period of despair. Her monastic creativity possibly even models strategies for our own survival in increasingly imperiled landscapes. In *The Great Derangement: Climate Change and the Unthinkable*, author Amitav Ghosh contrasts the Paris Climate Agreement's dry, factual tone with Pope Francis's 2015 encyclical on climate change, ultimately making a case for the necessity of artful language that touches readers on moral and spiritual levels in our efforts to preserve the earth for future generations.

*

During my conversation with Christine Whittington, the historian posted at the Matchless Mine, she mentioned that an important component of mine reclamation involves the planting of varietals such as western yarrow, lupine, and goldenrod. Someday I hope to revisit the Matchless, this time with a ceremonial offering of food, blank sheets of paper, and yarrow—a botanical long associated with healing—instead of my original morbid curiosity about Elizabeth Tabor.

*

Beyond the Matchless Mine shack, the closest approximation I've witnessed to nineteenth-century life in Leadville or other gold rush towns was a miner's shanty on the edge of the Rio Tigre, in Costa Rica. We passed it on another daylong excursion with Enrique, just before crossing a metal footbridge over the river. It was nothing more than an open-air shelter cobbled together from scrap wood and pallets. A dirt floor, with rusty pieces of corrugated steel for a roof. A few kitchen utensils on a stump and some kind of primitive cookstove. A robust middle class exists in Costa Rica; most *ticos* own their own property and modest houses, but this dwelling emanated despair and desperation.

Following Enrique's lead, we hiked up a jeep road thick with clay-rich mud, sometimes up to our ankles. This was *veranillo* season—the "little summer" in July that briefly interrupts the longer rainy season from April to November. The higher we climbed, the more sultry the air felt. The mud only grew thicker. When we finally reached the plateau summit, Enrique pointed out a pile of anteater scat and a line of small, half-formed cat tracks in the mud.

"An ocelot," he said, kneeling down before the tracks and beaming. "It was just here last night."

We told him a rumor we'd heard, that authorities recently caught a woman smuggling a live ocelot in a duffel bag aboard one of the commuter flights out of Puerto Jimenez.

"It wouldn't surprise me," he said. He went on to explain that the tracks we'd just seen belonged to an ocelot that had strayed beyond the boundaries of Corcovado and was therefore in serious danger. "Animals don't know borders or boundaries," he told us. "And young people in this area need more education about protecting the land."

Our attention turned skyward as we trudged farther up the plateau. We spotted a striking golden-hooded tanager, a relatively common species for which Enrique expressed his infinite enthusiasm. Then he coaxed out a black-hooded antshrike—a bird I didn't even know existed—from the underbrush.

Just after we turned around to head back down, Enrique stopped short. He held his hand up in that characteristic gesture that meant *pay attention.*

"You hear that?" he asked. To me it registered as nothing more than an ambient sound of the wind, but Enrique recognized it as a monsoon ghosting in off the gulf, headed directly toward us.

"We need to get off the plateau," he said.

Usually I'm the one to chide Lisa Mae for not bringing a raincoat on hiking trips, no matter how fair the weather appears. In this case she did bring a Gore-Tex shell, while I got soaked.

The rain eased off halfway down the hill, where we bumped into another wilderness guide and old acquaintance of

Enrique's. He was out following up on reports of a three-meter-long boa constrictor in an adjacent citrus orchard. Enrique decided to join him and suggested we wait near the jeep road. We joked nervously about the snake as Enrique disappeared down into a ravine. What if it decided to slither over in our direction? And how does one fend off a nine-foot constrictor?

Enrique reemerged into the orchard after several anxious minutes; he carried a very long wooden pole with a y-shaped tip.

"Oh my god," Lisa Mae said. "It that his snake-fighting pole?" We watched with rapt attention as he carefully balanced the pole with both hands and hoisted it high into a fruit tree. Using the forked tip he plucked out three oranges, then walked them over to us.

"That snake's long gone," he told us, while using his machete to slice the rind off the tip of the oranges with surgical precision. "*Salud*," he said, handing us each a perfectly sculpted, natural container from which to squeeze and savor the fresh juice.

We made one last stop that day, at a village farther down the Rio Tigre. From a guardrail-less bridge, we watched as a young couple sifted with shovels through a maze of man-made channels in the river below, searching for gold. In their early twenties, they were attractive and fashionable in a casual way, like students I might pass by on the college campus where I teach. They both wore knee-high Wellies; the young woman had on a black-and-white-striped top and neon-pink shorts. The memory of her brings to mind a young Elizabeth Tabor, hard at work on a mining claim with her first husband. Neither of these young *ticos* seemed particularly downtrodden or dispirited, just fully occupied in their tasks.

As we watched, we noticed an older, disheveled-looking man ambling up the bridge toward us. He smoked and talked to himself as he walked, then ogled Lisa Mae in a way that made me uncomfortable. He spoke briefly with Enrique, his breath rank with alcohol. Back in the car, Enrique told us that the man, a former orero himself, had long ago "gone crazy" with drink. And that he empathizes with this person's family.

"They're good people," he said, "but they can no longer handle him."

Unsettled by the encounter, I noticed myself initially judging the man as a casualty of his own choices, perhaps in the same way I first judged Elizabeth Tabor. But doesn't this explanation ignore the history of imperialism, extractive industry, and perhaps generational poverty that shaped the patterns of his and his family's existences? Larger forces constellated his life's trajectory.

*

In the winter of 2014, a surgeon performed a lumpectomy to remove an early-stage cancerous growth from Lisa Mae's breast. I helped care for her during the first few days of post-op; afterward she flew down to a Hindu ashram in a mountainous region of Costa Rica's Nicoya Peninsula. The ashram doubles as a working farm, with dairy cows and vegetable crops, but the monks wouldn't accept her help with the ever-present chores while she healed. Instead she spent two weeks wandering around, peering up into the canopy, her recovery aided by the sight and sound of turquoise-browed motmots, prothonotary warblers, crowned woodnymphs, magenta-throated woodstars.

After I finished up some work and flew down to Costa Rica, I was relieved to find her recovering so well, though I initially felt awkward at the ashram, surrounded by folks in white dhotis or monastics in orange robes, performing rituals that seem arcane to outsiders like me. Occasional bathing of two small deity statues with clarified butter and yogurt, for one. I also noticed the exceptional kindness they showed guests and one another. We took meals sitting on a clean cement floor near the kitchen, where a queer, heavily tattooed, former punk rocker turned monastic cooked flavorful curries, chutneys, and fresh-baked parathas, and afterward always made sure everyone enjoyed seconds and thirds. His soft-spoken gentleness reminded me of the warm-handed Episcopal minister from my childhood.

The leader of this particular ashram is one of few gurus in the Vaishnava tradition to initiate queer folks into monastic life—one of many reasons we admire him. His instruction can feel esoteric at times; he's translated the Bhagavad Gita, among many other texts, and has near-encyclopedic knowledge of Vedanta, but the idea that we should move away from a life of *always taking* remains at the core of his teaching. Instead he encourages us to exist for love—the same love he and his community lavish on their milk cows, the land, and all visitors.

At the ashram we were blessed with the presence of another writer: our friend Dana, who daylights as an ordained Southern Baptist minister back in North Carolina. She's married to a practicing Hindu, and wrote a book about their cross-religious relationship entitled *Saffron Cross*. We took long walks together, chatting about how strange and sweet this all was, and I noticed an unexpected peacefulness and quiet begin to wash over me. In the afternoons, after lunch and dialogues with the teacher, we hung out with our Finnish cartoonist friends, swimming together in a small, monk-made pond underneath a nearby waterfall, while they regaled us with their various bicycling ventures around the world—from Alaska to Northern California, across the entire continental United States, from Finland across western Russia.

One afternoon I spent half an hour watching a quartet of long-tailed manakins—the plump black males with bright-red crowns, sky-blue capes, and antennae-like tail feathers—singing in unison and dancing playfully on a dead tree snag just a few feet above my head.

*

The next week, Lisa Mae and I flew south, heading toward a national park named Amistad in honor of the mutual agreement between Panama and Costa Rica that sealed its creation and ongoing preservation. On the way up into the foothills we stopped at a botanical research station, where we lucked upon an abundant cluster of tanagers: speckled, bay-headed, golden-hooded, and silver-throated. Still one of my all-time favorite species, when I spotted them in my binoculars, like electric green and red jewels up in the bright canopy, I felt the same exhilaration as when Gabe showed me that first western tanager in his family's horse pasture, back in Colorado. Nothing else enlivens and transports my consciousness in quite the same way.

As lovely as the tanagers were, our sights were ultimately set elsewhere. The aspirational goal for the week was to find a rare and extraordinary bird known as the *resplendent quetzal*.

Quetzals dwell at higher elevations, so our local guide for this trip, Jaison, booked us a home stay at a working ranch on the edge of Amistad. The road up to the lodge was essentially a river of mud. Halfway up we had to stop for Jaison and the driver to install chains on the four-wheel-drive truck—something I'd never seen outside the context of a snowstorm. It still took multiple attempts to ascend the steepest stretch of road, the driver nearly giving up in frustration at several points.

With the ranch house as home base, we trudged around in damp pastures, our feet soaking wet as we searched for quetzals in between rainstorms. They eluded us for two entire days, but Jaison did help us spot an uncommon white-bellied mountain-gem hummingbird, American redstarts, blackburnian warblers, and a flame-colored tanager.

On our final morning near Amistad, Jaison led us higher into the forested foothills and resorted to playing quetzal calls from his cell phone in an attempt to lure one from the jungle. With our departure time approaching, we kept checking our watches in between longing gazes at the seemingly vacant canopy.

It felt like another fruitless search; we resigned ourselves to not finding what we'd hoped for. "Next time," we told one another, trying to put a brighter face on things.

Then, just as we turned back toward the lodge to pack up, a single, brilliant quetzal materialized from the denser forest and alighted on the limb of a wild avocado tree. Now that it was out in the open, it seemed pleased to show off its plumage—the dark-cherry belly below an emerald chest plate and greenish-blue cape. The proud face and round head topped with a shimmering green mohawk. Most remarkable were the long, emerald feathers ribboning down past the tail. In flight these feathers—almost three times the length of its body—undulated behind it like a marvelous Japanese kite or a sea serpent's tail.

We stood admiring the quetzal, feeling giddy with enchantment, until Jaison reminded us our taxi was waiting.

*

Back down on the Osa Peninsula, our attention shifted toward the ocean. Even while recovering from the lumpectomy, Lisa Mae surfed twice a day.

Blissed out from the sea and taking the quetzal sighting as an auspicious omen, we decided, after almost a decade of cohabitation, to get married. To celebrate the engagement our friends circled around us and made heart-shaped designs in the sand with stones burnished by the sea.

*

Along with rescuing the Canadian couple from Corcovado and an overloaded airplane, Enrique has participated in a handful of snake-bite rescues.

"Everyone comes together when these things happen," he tells us, referring to the potentially deadly bites from fer-de-lances that aren't entirely uncommon in Corcovado and the surrounding wilderness. He explains that, once the emergency call goes out, every available community member drops whatever they're doing to come to the aid of the injured. They work in shifts, and emergency litters fashioned from rope or even hammocks make the carrying slightly less arduous. But still it can take days to extract people from Corcovado's most challenging terrain.

During height of the COVID pandemic, I noticed myself returning, again and again, to the image of entire communities working to transport a snake-bite victim to safety. Shortly after retiring from her successful career as a nurse practitioner, and after surviving a serious bout with breast cancer herself, my mother volunteered to administer COVID vaccination shots. It embarrasses me to recall how, inspired by the first-wave men's movement and books like *Iron John*, I intentionally distanced myself from her in my misguided search for "heroes" as a young man.

If anything, COVID has taught us how much our survival depends on nurses, health care workers, and larger communities of care. I'm still leery of ascribing hero status to anyone, but we need nurses much more urgently than we need "wild men" living solipsistic existences.

I have equal admiration for Lisa Mae, a cancer survivor who performs acupuncture to heal her patients' chronic pain issues; she also raises herbs in our garden and concocts them into medicinal salves and potions. I'm done with hagiography but I deeply respect healers of all stripes.

Might we even embrace the concept of artists and writers as caretakers—or even healers—of a sort?

Though as Lisa Mae will be the first to tell you, healing is a complicated and imperfect science.

I will always live with complex PTSD, intermittent depression, and anxiety. These are rarely ailments from which anyone is ever entirely "cured."

Lisa Mae and my marriage proposal, for example, had the quality of a fairy-tale happy ending for us, but only six months after the actual wedding the 2016 episode of major depression nearly took us both out.

The absence of any panacea is especially true when it comes to the healing of lands, or of what the mining industry refers to as "reclamation."

Many environmentalists view the reclamation of formerly mined sites as a dubious prospect. A joke, even. It often entails the planting of some grass seed, and ever-rotating crews of subcontractors with unclear manifests. And what of the human communities that mining inevitably disrupts? If and when the Rio Tinto corporation opens a pit mine to extract copper from Arizona's Oak Flat—a site held sacred by the Apache Nation—will the wounds inflicted on local people ever fully heal?

In a more artistic sense, Elizabeth Tabor's own expressive writing was very often a direct attempt to reclaim her narrative from the more exploitative and often false stories propagated about her by newspapers, biographies, and a popular Hollywood movie named *Silver Dollar*. Unfortunately her attempts were largely futile and remain so—most who visit her homestead come away with only a cursory understanding of her life as "Baby Doe."

Another important element of my own ongoing recovery involved working with a trusted counselor who specializes in narrative therapy, a modality that helps one move beyond clinging to a single, monolithic story about ourselves or others—for instance that we're broken or failed or fallen—and to instead cultivate the multistoried aspects of our identities. Though creatively dilating our identities (and our literature) with a vibrant multiplicity of narratives doesn't mean everyone else will grant us the same courtesy.

*

You can't un-mine a mountain, the same way you can't un-exploit an oppressed worker or an innocent child. Though I like to imagine the phenomenon of un-mining—gold and silver winging off our fingers, helicoptering from earlobes and molars, melting off capitol rotundas—Tom's Baby levitating out of the Denver Museum of Nature and Science—all of it plummeting back into mine shafts, packing itself back into veins, the jackhammers operating in reverse to backfill all those chambers and stopes, rock by rock, boulder by boulder—fuses sparking away from bundles of dynamite, all the sluice water and arsenic surging backward up those old canvas hoses. Entire mountains and hillsides reconstituting themselves, like the way I sometimes stare in the mirror and imagine my younger, less scarred and weathered self.

Whimsical fantasies, both of them. Just as so many of us thrive in bodies with chronic illness or without certain abilities, though, we will continue finding ways to create meaningful lives, mosaic identities, and networks of care on a radically altered planet. It may look grim at times, but the future is still ours to dream.

*

In another of my most memorable bird-watching moments, I once stood still for nearly an hour on the edge of a Costa Rican cow pasture, sweeping the distant jungle canopy with my binoculars, hoping to spot a vulnerable bird species that I'd heard about but never actually seen. After thirty or forty minutes of silent scanning, my feet and neck began to complain. The midmorning sun ovened the air up toward ninety. My stomach and exposed skin craved breakfast in the shade. Until, high up on the dark wall of rainforest, a tiny bright doorway of color flashed in the left corner of my binocular lens. Snapping back to the left and fine-tuning the magnification, it gradually came into focus: a turquoise cotinga—an exquisite, voltaic-blue bird with a heart-shaped patch on the chest in regal purple. Did I mark it down in my notebook? Probably. I'd like to think I also bowed to it, thanked it for revealing itself and for persevering as one of only five or ten thousand of its kind left on the planet.

*

With nine-foot surfboards tucked under our arms, Lisa Mae and I sprint, awkwardly, across the sunbaked sand. When we reach the ocean's foamy edge the salt water feels like balm on the stinging soles of our feet. We attach our leashes and exchange glowing looks before the long paddle out to our favorite Costa Rican break.

The water's deliciously cool on our feet, but once I fully submerge it feels almost *too warm*, like thermally heated spring water—especially compared with Oregon, where ocean temperatures fluctuate around fifty-five degrees.

We slot ourselves into the channel, waves welling up on either side of us as we continue our push. Just before collapsing into white, the sun flashes platinum on each swell's wind-textured skin.

The taste of salt brine on our lips. The mild sting of it on our faces and in our eyes.

Bobbing out beyond the break, giving our chest and arm muscles a rest, we peer across to the long peninsula on the south side of the Golfo Dulce. Shrouded in haze on the distant horizon, the long finger of land appears almost mystical. It seems to stretch on forever until it reaches the very tip of Panama.

In the past we've witnessed humpbacks breaching out here. We've delighted at the site of dolphin pods and the occasional turtle.

Way up above, frigate birds spiral with their long, retracted wings. Closer to the shore, swallows and martins hunt for insects in tight figure eights.

It's 2019 and I'm grateful to have my health and vitality back. To no longer endure anhedonia—the inability to enjoy *anything*.

Now, just being out here in the water with Lisa Mae, watching her catch some fun rides with her radiant, wide-open smile, is more than enough.

I keep my eyes peeled for just the right bump on the ocean's surface.

I notice I'm also looking out for the tour boat full of interlopers, hoping they don't show up and ruin our day.

After paddling myself farther out into position, I strike up a promising rapport with an incoming swell. A curving, kinetic sculpture of milky blue-green water. While I don't get quite the exciting drop I hoped for, it does grant me a long frontside ride with plenty of room to maneuver my board up and down the face. Once the left shoulder peters out, I redirect back toward the power source. With my heels now facing the wave as it tightens up, I squat down and slingshot myself into the cove, fingers dragging in the curl behind me, riding it as far as I can before the long paddle back.

Having lived for a period devoid of even the most basic state of contentment, I want to cling to this exalted feeling forever.

I scan the beach for a nearby property owner, a leathery-skinned Brit in his late fifties who surfs much better and more aggressively than I do. I hope to carve out what I can before he shows up again, as I don't enjoy surfing with him.

Then again, if I was a beginner surfer, would I enjoy being out in the water with me? Have I dominated waves, even from my friends? From my own partner?

It's easy to critique acts of colonization and greed when sitting alone at a computer. The endeavor grows thornier when someone takes a wave that, according to standard surfer etiquette, rightfully belongs to you. We forgive beginner surfers their transgressions. But once in Oregon a younger surfer stole my wave on purpose—despite my forcefully calling out my line—and then laughed with his buddies about it. Back on the beach, I felt primed for a fist fight—an all-too-common occurrence in the territorial culture of surfing. Fortunately my friends helped me cool down.

Even here in Costa Rica, with literally no one else around, I notice my mind hissing with competition for limited resources, with desire to hollow out every surfable possibility from the ocean, to *get it while I can* before the bonanza dies down.

On the paddle back from my memorable ride, a set of crashing waves slows me down—what's known in surf parlance as a *cleanup set*. They wash over me, rinse out my sinuses, hold me under for a moment in the slightly cooler, subsurface water. Emerging from the fizz, I catch a breath of humid air and aim my board once again out toward the oncoming swells.

I'm not consciously aware of it at the time, but from the vantage point of my writing desk almost exactly five years into the future—where I live with the aching knowledge that this particular Osa trip was quite possibly our last, after COVID ended our yearly retreats and most of our friends sold their properties to much wealthier owners— I can hear it so clearly:

These breaking waves with their lacy white fingers, typing out their endless, sibilant verse—a language that no written sentence can ever hope to replicate.

Whispering while they type, urging my mind to *hush*.

Acknowledgments

I began drafting elements of *A Field Guide to the Subterranean* in 2013; over the years it evolved through multiple iterations with support, contributions, and sustenance from many readers, friends, family members, editors, research sources, organizations, landscapes, and waterways.

I'm grateful to my late father, Roger Hocking, and my late friend and mentor, Barry Lopez; this book is dedicated to them both. I also wish to thank my mother, Harrel Lawrence, and my stepfather, Jerry McMahan, for their love and hospitality during research trips in Colorado, and family members Stephanie Zehren, Nat Hocking, Jenny Powers, Tracy Powers, and all my McMahan siblings and relatives.

Endless thanks to my wonderful editor, Dan Smetanka, for his subtle, astute edits and guidance; the text benefited immensely from his input. And to my agent and friend, Matt McGowan, by my side for well over a decade now, without whom this project may not have seen the light of day. I'm grateful to Megan Fishmann, for her energetic publicity work and shared love for *Point Break* that helped parachute this book into the world with *100 percent pure adrenaline*; Farjana Yasmin and Nicole Caputo, for the sublime cover design and art direction; Laura Berry, for the artful typesetting and interior book design; Madelyn Lindquist, for the thoughtful copyedits; and Ashley Kiedrowski, for her digital marketing work. Thanks also to Rachel Fershleiser, Yukiko Tominaga, and the entire team at Counterpoint Press.

Acknowledgments

The Regional Arts and Culture Council (RACC) provided grant funding that supported this project in its nascent stages. With RACC support I produced an earlier, encapsulated version of the narrative in the form of a twenty-two-page chapbook entitled *Reclamation: Essays*, with cover design, artwork, and letterpress printing by Caitlin Harris from Wheelhouse Press and interior text design by Brian David Smith. The Sou'wester Foundation provided valuable time and space for multiple writing residencies; thanks also to Krissy Evans, for offering me a house rental on the Oregon Coast, where I completed the first full manuscript draft, and to Annie Rector, for sharing her home during a research and writing trip in Glenwood Springs, Colorado.

I'm grateful to Chad Perman at *Bright Wall/Dark Room* for publishing an earlier version of "Groundshock." And to Molly, Liza, and Paul from *Big Big Wednesday*, for publishing a previous version of "Origins." Thanks to Mike Midlo and Shannon McNerny for hosting me at a 2016 Fishtrap Summer Writers Conference, where I shared excerpts from an early draft of the "Outburst" sections. Robin Bacior from *Willamette Week* also provided invaluable support, as did Joshua James Amberson and other *Portland Mercury* writers and editors.

Extra-special thanks and a lifetime of good vibes to Ann Petriolunas, for the close read and helpful suggestions. Also to Tessa Torgeson, A. M. O'Malley, Gabriel Urza, Alicia Jo Robbins, Andreas Trolf, Paul Thompson, Amy Knauer, and others for early manuscript reads and feedback.

Thanks to members of our writing group: Jessie Page, Michael Heald, Evan P. Schneider, Martha Grover, Erin Ergenbright, Kjerstin Johnson, and Jessica Yen. Many thanks also to Héctor Tobar, Erica Berry, Rebecca Skloot, Steven Church, John Calderazzo, Steven Schwartz, Deborah Gwartney, Aaron Gilbreath, Kevin Sampsell, B. Erin Cole, Nicole Walker, Margot Singer, Brett Anthony Johnston, Kyle Boelte, Karen Karbo, Cheryl Strayed, Cliff Ponca, Liz Asch, Karleigh Frisbie Brogan, Josh Gaines, Andrew Barton, Dao Strom, Trevor Spangle, Erika Bolstad, Melissa Faliveno, Amy Wheeler Harwood, Noah Shannon, Lawrence Siulagi, Andy Weiss, Allison Alter, John and Karla Baise, John and Jess Skibo, and many others.

My project of unearthing the past may not have reached the surface without a Corporeal Writing workshop with Lidia Yuknavitch and Domi Shoemaker, who helped me discover light sources while stumbling around in the underworld. Love to you both and thanks for all you do for the Portland writing community and beyond.

I owe so much to my supportive and brilliant colleagues at Portland State University: Paul Collins, Leni Zumas, Janice Lee, John Beer, Gabriel Urza, Consuelo Wise, Marie Lo, Liz Ceppi, Walidah Imarisha, Jennifer Kerns, Kathi Inman Berens, Julie Perini, and many others. And to all our exceptionally talented PSU writing students, who inspire me every day.

I owe a major debt of gratitude to Raewyn W. O'Connell, for her book *Masculinities*, and to Judy Nolte Temple, for her illuminating and meticulously researched book *Baby Doe Tabor: The Madwoman in the Cabin*, which helped me understand Elizabeth Tabor in a new and brighter light. This book came about in part thanks to my lifelong fascination with the Golden Nature Guide *Rocks and Minerals* and the Golden Field Guide *Birds of North America*. I'm also particularly indebted to the Southern Ute Indian Tribe;

their online timeline of historical events that occurred on their homelands provided key references and dates.

I'm grateful to the authors and publishers of the following works, all of which provided vital inspiration and insights: *Glenwood Springs: The History of a Rocky Mountain Resort* by Jim Nelson; *An Indigenous Peoples' History of the United States* by Roxanne Dunbar-Ortiz; *Arctic Dreams* and *Horizon* by Barry Lopez; *Meet Joe Copper: Masculinity and Race on Montana's WWII Home Front* by Matthew L. Basso; *White Magic: Essays* by Elissa Washuta; *Confessions* by Saint Augustine; *Natural History* by Pliny the Elder; *The Anthropology of Turquoise* by Ellen Meloy; *Annals of the Former World* by John McPhee; *A Brief History of Earth* by Andrew H. Knoll; *Angle of Repose* by Wallace Stegner; *The Faraway Nearby* by Rebecca Solnit; *Bluets* by Maggie Nelson; *The Gnostic Gospels* by Elaine Pagels; *Leaves of Grass* by Walt Whitman; *The Elusive Dream* by Anna Johnson and Kathleen Yajko; *Leadville: A Miner's Epic* by Stephen M. Voynick; *The Lincoln Memorial and American Life* by Christopher A. Thomas; *The Shining* by Steven King; *Cosmos* by Carl Sagan; *The Hotel New Hampshire* by John Irving; *Iron John* and *The Sibling Society* by Robert Bly; *The Book of Delights* by Ross Gay; *Braiding Sweetgrass* by Robin Wall Kimmerer; *Ongoingness* by Sarah Manguso; *The Noonday Demon: An Atlas of Depression* by Andrew Solomon; *Saffron Cross* by J. Dana Trent; *The Great Derangement: Climate Change and the Unthinkable* by Amitav Ghosh; and *Oak Flat: A Fight for Sacred Land in the American West* by Lauren Redniss.

Thanks also to authors Susan Sontag, David Treuer, Walidah Imarisha, Alexis Pauline Gumbs, Joseph Campbell, A. O. Scott, Michiko Kakutani, and Lewis Mumford, and filmmakers Alain Resnais, Jean-Luc Godard, Stanley Kubrick, Rodney Ascher, and Paul Schrader. And to historian Christine Whittington, for her generosity with her time at the Matchless Mine, as well as Adrian Brown, for driving me to safety. My research on various aspects of Colorado state history relied on the following media sources: Colorado Historic Newspapers Collection, *The Aspen Times*, *The Grand Junction Daily Sentinel*, *The Denver Post*, *Westword*, and others. For historical information about the Ludlow Massacre, I'm grateful to PBS's *American Experience*, George Mason University's "History Matters" archives of *New York Times* coverage, the United Mine Workers website, and Woody Guthrie's song "Ludlow Massacre." Thanks also to *Harper's Magazine* and their publication of Barry Lopez's "Sliver of Sky" essay, and to the *Rabbit Hole* podcast. The following organizations were immensely helpful with my research: the Glenwood Springs Frontier Historical Museum, the Denver Museum of Nature and Science, High Country Gems and Minerals, and the National Mining Hall of Fame and Museum.

I will always appreciate the Colorado Outward Bound School, my COBS instructors, and all my fellow students during our Leadership Mountaineering course, as well as everyone I worked with during my days as a backpacking guide in the San Juans: the Colvig family, Shari, Laura, Jeff, Liz, Luis, Teddy, Sarah, Erica, and so many others. Deep thanks to La Plata County and their exceptional Search and Rescue efforts.

Many thanks to our friends in Costa Rica, most especially Felipe Arayas, as well as Monica Bauer, Eddy and Michelle, Jaison, and many others. And to the following people for their kindness and wisdom: Swami Tripurari,

Gaurasundara Dasa, Kamalaksa, Krisangi, Gurunishta Das, Gauravani, and J. Dana Trent.

Thank you to Peter Bennett for the narrative support and thoughtful questions; the late Jean Amir and his family; A.J., Jamie, and our Stronger Skatepark crew; Harper and everyone at the Independent Publishing Resource Center; and photographer K. B. Dixon.

There will never be enough thanks for Erica Simpson, her parents, Bruce and Adrian Staff, and anyone else who endured my men's movement phase in the 1990s.

Special thanks to T.T., for our ongoing text exchange that has been a genuine life support at so many times over the past decade. And to my Golden Tiles bandmates and dear friends, Joshua James Amberson and Oliver Stafford, for brightening up my Wednesday nights.

Deep gratitude to Gabriel Liston, for the careful manuscript read and suggestions, and for his lifelong friendship. And to Sheila, Ron, Caleb, Solomon, Rachel, and Levi—my surrogate family growing up, who always made space for me at the Mountain House.

Thank you to my faithful research assistant and friend, Bo, for inspiring me to always seek light and warmth.

Finally—and most of all—to my first reader, editor, chief strategist, and best friend, Lisa Mae Osborn, for the supercharged support and encouragement for this project and all my creative endeavors. I love you and couldn't do any of this without you.

A portion of the proceeds from the sale of this book will be donated to the Portland State University Native American Student and Community Center.

JUSTIN HOCKING is the author of the *The Great Floodgates of the Wonderworld: A Memoir*, which won the Oregon Book Award and was a finalist for the PEN Center USA Award for Nonfiction. He served as the Executive Director of the Independent Publishing Resource Center (IPRC) from 2006 to 2014 and is a recipient of the Willamette Writers Humanitarian Award for his work in writing, publishing, and literary outreach. He teaches creative writing in the MFA and BFA programs at Portland State University. Find out more at justinhocking.net.